Gait and Balance Control in Typical and Special Individuals

Gait and Balance Control in Typical and Special Individuals

Guest Editor

Luis Augusto Teixeira

MDPI

Basel • Beijing • Wuhan • Barcelona • Belgrade • Novi Sad • Cluj • Manchester

Guest Editor
Luis Augusto Teixeira
Human Motor Systems
Laboratory
University of São Paulo
São Paulo
Brazil

Editorial Office
MDPI AG
Grosspeteranlage 5
4052 Basel, Switzerland

This is a reprint of the Special Issue, published open access by the journal *Biomechanics* (ISSN 2673-7078), freely accessible at: https://www.mdpi.com/journal/biomechanics/special_issues/32743M3HV8.

For citation purposes, cite each article independently as indicated on the article page online and as indicated below:

Lastname, A.A.; Lastname, B.B. Article Title. *Journal Name* **Year**, *Volume Number*, Page Range.

ISBN 978-3-7258-5255-0 (Hbk)
ISBN 978-3-7258-5256-7 (PDF)
https://doi.org/10.3390/books978-3-7258-5256-7

Contents

About the Editor

Luis Augusto Teixeira

Luis Augusto Teixeira is an Associate Professor and Head of the Human Motor Systems Laboratory at the School of Physical Education and Sport, the University of São Paulo, Brazil. His research focuses on understanding balance control in challenging situations, particularly how the body executes fast, reactive responses to regain stability from predictable or unpredictable perturbations, thereby preventing falls. More recently, he has been evaluating how the principles of balance control and motor learning can be applied to training activities. This work aims to improve voluntary dynamic balance and the speed of reactions to situations that can potentially induce a loss of balance, specifically in older adults and individuals with neurological conditions. His research interests encompass the intricate connection between motor, physical, and cognitive domains across different stages of aging.

Preface

The field of neuromechanics stands at the intersection of neuroscience, engineering, and physiology, aiming to unravel the complex control systems that govern human movement. At its heart lies "gait and balance control", something we often take for granted but is an essential aspect of life, from an infant's first steps to an elite gymnast's flawless routine. Its disruption can profoundly impact an individual's independence and quality of life.

This Reprint of the Special Issue of *Biomechanics* titled "Gait and Balance Control in Typical and Special Individuals" aims to provide a comprehensive and up-to-date overview of the latest advancements in this field. Our purpose is to synthesize recent findings and demonstrate how sophisticated biomechanical techniques are revolutionizing our understanding of both typical and pathological movement. We seek to highlight not only the intricacies of normal gait but also the unique challenges faced by a diverse array of populations, from young children to the elderly and those with chronic health conditions.

The motivation behind this work is rooted in the persistent and evolving questions that define this area of research. We aimed to consolidate the disparate but impactful studies that have emerged recently, providing a unified platform for research in this field. As researchers, we recognized the need to move beyond isolated findings to build a cohesive narrative that underscores the clinical and societal importance of studies on gait and balance. The insights within this Special Issue—from understanding developmental variability and athletic performance to addressing the profound effects of neurologic malfunctions such as cerebral stroke and Parkinson's disease—underscore the real-world implications of our work.

This Special Issue is primarily addressed to a broad scientific and clinical audience, serving as a crucial resource for researchers in biomechanics, neuroscience, rehabilitation science, and sports medicine who wish to stay at the forefront of the field. Additionally, this Reprint is intended for clinicians, including physical therapists and occupational therapists, who can apply the reported findings to improve their patient assessment or intervention strategies. Moreover, we believe this collection of papers will be of great interest to students and academics who are new to the field, offering a curated entry point into the most pressing and dynamic topics in contemporary neuromechanics. We hope to inform and inspire future generations of researchers to continue exploring the dynamic complexities of human movement.

Luis Augusto Teixeira
Guest Editor

biomechanics

MDPI

Article

Nonlinear Gait Variability Increases with Age in Children from 2–10 Years Old

Bryon C. Applequist [1,*], Zachary L. Motz [2] and Anastasia Kyvelidou [3]

1 Department of Kinesiology, Texas A&M University-Corpus Christi, Corpus Christi, TX 78412, USA
2 Department of Biomechanics, University of Nebraska at Omaha, Omaha, NE 68182, USA; zmotz@unomaha.edu
3 Department of Physical Therapy, Creighton University, Omaha, NE 68178, USA; anastasiakyvelidou@creighton.edu
* Correspondence: bryon.applequist@tamucc.edu

Abstract: Background: Linear methods of analysis of variability are concerned with the magnitude of variability and often consider deviations from a central mean as errors. The utilization of nonlinear tools to examine variability allows for the exploration and measurement of the patterns of variability displayed by the system. This methodology explores the deterministic properties of biological signals, in this case, gait, or how previous iterations within the gait cycle influence subsequent and future iterations. The nonlinear analysis of gait variability of the joint angle time series has not been investigated in developing children. **Methods**: We collected 3 min of treadmill walking data for 28 children between the ages of 2 and 10 years old and analyzed their joint angle time series using nonlinear methods of analysis (sample entropy, largest Lyapunov exponent, and recurrence quantification analysis). **Results**: Our results indicate that the nonlinear variability of children's gait increases as children age. Interestingly, this contrasts with the findings from our previous work that showed a decrease in linear variability as children age. The combination of a decrease in linear variability, or a refined and improved stability of gait, as well as an increase in nonlinear variability, or an increase in the sophistication and quality of movement patterns, suggest an overall maturation of the neuromuscular system. **Conclusions**: Our study indicate that there is a refining of gait with age and motor maturation. This refining speaks to the overall multifaceted organization of systems that defines the maturation of gait.

Keywords: biomechanics; variability; gait; nonlinear; children

1. Introduction

Researchers interested in the movement sciences have used linear methods to analyze movement variability. When linear tools are utilized, the magnitude of variability becomes the emphasis, while assuming that all repetitions of a behavior, such as gait cycles, are independent from what has happened before or what will happen after [1]. Another way to view variability in the world of movement science is through the lens of nonlinear methods. Nonlinear methods focus on the structure of variability by scrutinizing patterns in the variability across time. Nonlinear methods view the determinism within a series of movements or how one movement influences the next, and so on. Both the magnitude and structure of variability offer valuable information regarding movement and can differ between persons [1]. Interestingly, some movements may fall within the same magnitude of variability, but possess differences in the structure of the variability. This is a key distinction, as structure of variability has been associated with the health of biological systems. Healthy systems are those that possess a certain amount of stability but remain adaptable [2]. Both exceeding regularity across repetitions, with extremely periodic organization of variability, as well as an absence of consistency, with random organization of variability, have been linked to poor health [3,4]. The extrema of variability can be thought of as two ends of a spectrum, but in the middle lives a deterministic but non-periodic pattern that provides a balance between flexibility and stability of behavior. This middle state is associated with maximum complexity, which is defined as the highly variable fluctuation in physiological processes, resembling mathematical chaos [1].

The utilization of nonlinear tools allows for the exploration and measurement of the patterns of variability displayed by the system. These nonlinear tools can be used to study gait in humans [5–8]. Previous work investigating the progression and development of the gait of healthy children has focused mostly on the spatiotemporal, kinematic, and kinetic aspects of children's gait, with little focus on variability. Of the investigations into variability, there has been even less of a focus with an eye toward the nonlinear dynamics and nonlinear variability. One particular study showed that the variability measures of the spatiotemporal aspect of children's gait, as well as the nonlinear measures of the dynamics of gait in children, are age-dependent, and do not mature and become adult-like until after 10 years of age [9]. These results were in contradiction to the previously conceived notion that children's gait was mostly mature by the age of 4 years old [10]. In a previous study, we indicated a lack of early maturation of spatiotemporal measures as well, utilizing both linear and nonlinear methods [11]. The linear methods of variability (standard deviation and coefficient of variation) proved to be extremely age-dependent, with younger children exhibiting more variability than their older counterparts. Nonlinear measures also showed differences with age, as regularity (entropy) and complexity (detrended fluctuation analysis (DFA)) increased with age.

The use of entropy and DFA to analyze the spatiotemporal time series provided new insight into the developmental trajectories of children's gait, while paving the way for further investigation into different aspects of children's gait using nonlinear variability methods [9]. A specific aspect of gait that nonlinear methods have been successful at analyzing in adult gait is the joint angle time series during walking. Entropy, largest Lyapunov exponent (LyE), and recurrence quantification analysis (RQA) have all been used to analyze the joint angle time series of the lower extremity to describe gait variability and gait variability changes [12–15]. Specifically, analyzing the joint angle time series with nonlinear measures has enabled the detection of differences between adult walkers, with and without pathology [8,12,14].

Utilizing these nonlinear measures for the investigation of joint angle time series in children has not been examined. To further the understanding of children's gait and the development of children's gait variability, nonlinear measures of analysis should be utilized on the joint angle time series of children. Differences in the structure of variability of the joint angle time series of children at different points in their development should be identifiable using nonlinear tools of analysis. The use of nonlinear methods could shed light on the potential control mechanisms being used and the refinement of gait throughout development and will provide a launching point for the comparison of the natural trajectory of gait development. This new information can be used to help understand various types of pathological gait in children.

The purpose of this study was to assess the development of kinematic gait variability in children from ages 2–10 years old using nonlinear methods of analysis. To do this, we grouped children into four separate age groups consisting of 2–3-year-olds, 4–5-year-olds, 6–7-year-olds, and 8–10-year-olds. We then had them walk on the treadmill for three minutes. The joint angle time series of the lower extremity were then analyzed. We hypothesized that as children aged, their gait variability will become more regular and exhibit greater adaptability. We also hypothesized that with age, children will display less stride-to-stride fluctuation in their gait.

2. Materials and Methods

2.1. Subjects

Our study involved 28 boys and girls split into four separate age groups. The age groups consisted of 2–3-year-olds (n = 7), 4–5-year-olds (n = 7), 6–7-year-olds (n = 7), and 8–10-year-olds (n = 7) (Table 1). Power analysis was conducted to determine that groups of four subjects were necessary to achieve adequate power. All participants provided parental informed consent and child assent before any research activities commenced, as approved by the university's Institutional Review Board. Healthy children, free from any musculoskeletal disorders, injuries, or developmental delays, were included in our study.

Table 1. Subject demographics by age group for participants.

	2–3-Year-Olds (N = 7)	4–5-Year-Olds (N = 7)	6–7-Year-Olds (N = 7)	8–10-Year-Olds (N = 7)
Gender (male/female)	4/3	3/4	3/4	3/4
Age (months)	35.9 ± 7.3	58.57 ± 5.7	81.57 ± 6.37	115.6 ± 6.02
Body mass (kg)	13.67 ± 2.5	17.31 ± 1.4	25.99 ± 4.97	38.44 ± 5.23
Body height (m)	0.92 ± 0.08	1.04 ± 0.03	1.22 ± 0.06	1.38 ± 0.03
Onset of walking (months)	12.14 ± 0.69	12.57 ± 2.15	12.21 ± 1.30	13.29 ± 0.95
Walking speed (m)	0.56 ± 0.16	0.78 ± 0.14	0.92 ± 0.08	1.07 ± 0.11

Note: values are shown as mean ± standard deviation.

2.2. Experimental Procedures

All subjects were provided with tight-fitting athletic shorts to be worn during the data collection process to ensure accurate marker placement for the motion capture system. The subject's shoe size was then determined, and they were provided with a standard laboratory shoe (Nike Free 5.0). The lab-provided shoe was employed to eliminate potential differences in footwear styles worn by the children, while also providing a normalized control. The Nike Free 5.0 is considered a "minimalist" style shoe, which mostly mimics barefoot conditions [16]. Study participants were given time to familiarize themselves with the treadmill (Bertec Corp, Columbus, OH, USA) and the lab shoe, while a self-selected comfortable walking speed was determined. Previous work has shown that treadmill walking functions to reduce gait variability compared to overground walking [17]. However, this work analyzed spatiotemporal gait and not the variability of joint angle kinematics. Treadmill walking was selected in comparison to overground walking because of the requirement for a long time-series of unbroken data. Retro-reflective markers were then placed on the subject at specific anatomical locations of the foot, shank, thigh, and pelvis, according to the marker systems established by Nigg et al. [18] and Vaughan et al. [19]. Lower extremity marker locations were acquired for one three-minute trial per condition at 100 Hz using an eight-camera motion capture system (Vicon Motion Systems, Oxford, UK). The participants performed the walking trials positioned 2 m in front of a screen in a virtual reality environment. However, to simulate walking on a stationary treadmill with static optic flow stimulation, a picture of the static room surround was projected on the virtual reality screens. The two conditions consisted of at least 3 min of walking at a self-selected comfortable walking speed while barefoot and while wearing the lab-provided footwear. The data were left unfiltered so as not to affect or influence potential biological signals within the data. It has been shown that filtering the data can lead to altered nonlinear results [20].

2.3. Data Analysis

We computed the joint angle time series in the sagittal plane of the ankle, knee, and hip joints utilizing Visual 3D software (C-Motion Inc., Germantown, MD, USA). Data processing and analysis were conducted using custom Matlab scripts (The Mathworks Inc., Natick, MA, USA). Lower extremity sagittal plane joint angles were analyzed because most bipedal motion occurs in the sagittal plane during gait. Sample entropy (SE) was calculated to determine the organization of the gait variability of each joint angle time series. A lower value of SE alludes to more rigidity and regularity and thus, less variability in the time series. A larger value of SE means more variability in the time series. The structure of the gait variability during the walking trials was evaluated using the LyE. The methodology of the LyE has been outlined in great detail by Wurdeman et al. in a separate publication [13]. In brief, the LyE measures the exponential divergence of the movement trajectories within a reconstructed state space [21]. Recurrence quantification analysis (RQA) of the joint angle time was also performed. RQA is a method of nonlinear data analysis for the investigation of dynamical systems. RQA quantifies the number and duration of recurrences of a dynamical system presented by its phase space trajectory, and it has been proven to be a good way to analyze the predictability and complexity of the system. We evaluated the percent determinism (%Det) and mean line (MLine) for our data. %Det is the percentage of recurrent points forming line segments parallel to the main diagonal line. The presence of these lines reveals the existence of a deterministic structure. MLine is the average length of all the line segments on the RQA plot. The MLine is a good indicator of the predictability of the time series.

2.4. Statistical Analysis

A one-way ANOVA with four factors (four age groups) was performed to determine the statistical significance for each of the dependent variables for the ankle, knee, and hip joints angle time series, respectively. The dependent variables include SE, LyE, %Det, and MLine. When significant effects were determined, post hoc comparisons were performed using the Tukey method. Statistical analysis was completed in SPSS Statistics 29 (IBM Corporation, Armonk, NY, USA).

3. Results

Mean and standard deviations of all variables can be found in Table 2. Significance between the variables is denoted by symbols.

Table 2. Group means for sample entropy, largest Lyapunov exponent, and recurrence quantification analysis for the 2–3, 4–5, 6–7, and 8–10-year-old groups.

	2–3 (N = 7)	**4–5 (N = 7)**	**6–7 (N = 7)**	**8–10 (N = 7)**	**Sig.**
			Sample Entropy		
Ankle	0.322 ± 0.016	0.315 ± 0.037	0.308 ± 0.045	0.282 ± 0.048	
Hip	0.223 ± 0.025	0.226 ± 0.016	0.229 ± 0.016	0.195 ± 0.021	‖¶
Knee	0.230 ± 0.017	0.233 ± 0.036	0.232 ± 0.043	0.261 ± 0.053	
			Largest Lyapunov Exponent		
Ankle	1.06 ± 0.10	1.17 ± 0.10	1.28 ± 0.10	1.43 ± 0.08	†‡‖¶
Hip	0.60 ± 010	0.67 ± 0.13	0.75 ± 0.19	0.89 ± 0.15	‡‖
Knee	1.11 ± 0.08	1.19 ± 0.08	1.24 ± 0.10	1.30 ± 0.10	†‡

Table 2. *Cont.*

	2–3 (N = 7)	4–5 (N = 7)	6–7 (N = 7)	8–10 (N = 7)	Sig.
		Recurrence Quantification Analysis %Determinism			
Ankle	73.2 ± 9.9	78.8 ± 5.6	83.6 ± 6.2	86.9 ± 4.2	†‡
Hip	76.9 ± 6.7	80.9 ± 2.4	90.5 ± 4.9	95.7 ± 2.2	†‡§‖
Knee	71.5 ± 9.8	76.1 ± 13.5	77.5 ± 15.3	83.9 ± 5.3	
		Mean Line			
Ankle	2.51 ± 0.24	2.55 ± 0.44	2.60 ± 0.27	3.06 ± 0.40	‡
Hip	4.87 ± 1.75	5.31 ± 1.04	5.91 ± 1.41	6.10 ± 1.44	
Knee	2.10 ± 0.59	2.66 ± 0.59	2.60 ± 0.39	3.51 ± 1.21	‡

Note: values are shown as mean ± standard deviation. Special characters for the following represent a $p < 0.05$, significant differences between groups 2–3 and 4–5. † $p < 0.05$, significant differences between groups 2–3 and 6–7. ‡ $p < 0.05$, significant differences between groups 2–3 and 8–10. § $p < 0.05$, significant differences between groups 4–5 and 6–7. ‖ $p < 0.05$, significant differences between groups 4–5 and 8–10. ¶ $p < 0.05$, significant differences between groups 6–7 and 8–10.

To see if the results were age-dependent and not a function of biomechanical changes related to growth, we investigated the linear relationship between both age and leg length and age and gait speed. In the present study, both leg length (r = 0.966 $p < 0.001$) and gait speed (r = 0.839 $p < 0.001$) increased linearly with age. Thus, we also conducted comparisons while normalizing the dependent variables with respect to both leg length and gait speed.

3.1. Sample Entropy of Joint Angle Time Series

The results for the SE analysis are shown in Figure 1. There was only a significant effect of age for the SE of the hip joint time series F (3,24) = 4.296, $p = 0.015$. Specifically, post hoc comparisons revealed that the 8–10-year-old group exhibited significantly greater SE at the hip compared to both the 4–5-year-old group ($p = 0.04$) and the 6–7-year-old group ($p = 0.019$). The SE for the ankle and knee joint angle time series did not produce an effect ($p > 0.05$). There was a significant linear trend of age for SE at the hip (r = 0.394, $p < 0.05$), as well as at the knee (r = 0.481, $p < 0.05$), but not at the ankle ($p > 0.05$).

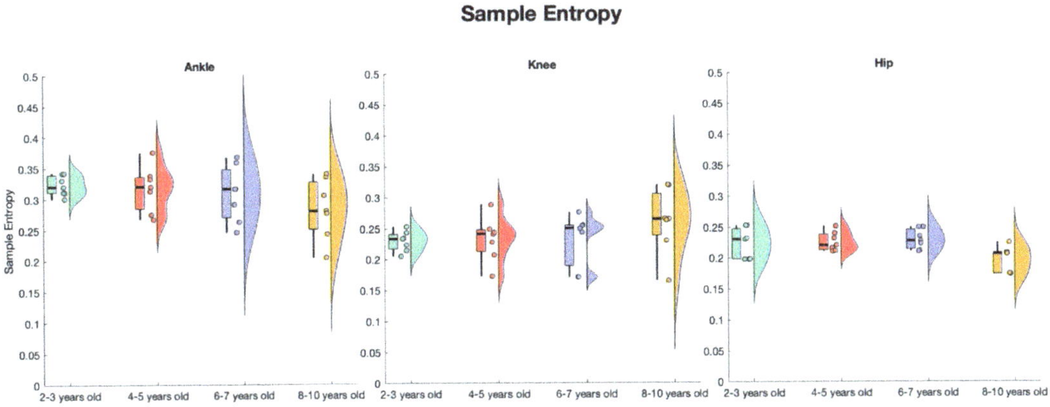

Figure 1. Violin and box plots showing the distribution of the sample entropy of the ankle, hip, and knee joint time series. Data are reported for the age groups.

Normalized comparisons for SE showed a significant effect at both the hip F (3,24) = 29.51, $p < 0.001$, and knee F (3,24) = 10.17, $p < 0.001$, joints. Post hoc comparisons at the hip joint showed that the normalized SE of the hip joint angle time series decreased with age. Specifically, the 2–3-year-old group showed the greatest SE, and it was significantly greater than that of all other age groups ($p < 0.05$). The 4–5-year-old group exhibited significantly greater SE than did the 6–7-year-old group and the 8–10-year-old group ($p < 0.05$). The results for the 6–7-year-old group were also significantly greater than for the 8–10-year-old group ($p < 0.05$). Post hoc comparisons at the knee joint showed that the normalized SE of the knee joint angle time series also significantly decreased with age, except for in the 6–7-year-old group and 8–10-year-old group ($p < 0.05$). Specifically, the 2–3-year-old group exhibited a significantly greater normalized SE at the knee compared to that of the 4–5-year-old, 6–7-year-old, and 8–10-year-old groups ($p < 0.05$). The 4–5-year-old group showed a significantly greater normalized SE at the knee than did the 6–7-year-old group and the 8–10-year-old group ($p < 0.05$), but there was no significant difference between the 6–7-year-old and the 8–10-year-old groups ($p > 0.05$).

3.2. Lyapunov Exponent of Joint Angle Time Series

The results for the LyE analysis are shown in Figure 2. There were significant effects of age group for the LyE of the ankle joint time series F (3,24) = 19.686, $p < 0.001$, the hip joint time series F (3,24) = 4.958, $p = 0.008$, and the knee joint time series F (3,24) = 6.151, $p = 0.003$. Post hoc comparisons at the ankle joint indicate that the 2–3-year-old group exhibited significantly lower LyE compared to that of the 6–7-year-old group ($p = 0.001$) and the 8–10-year-old group ($p < 0.001$). The 4–5-year-old group showed significantly lower LyE values compared to those of the 8–10-year-old group ($p < 0.001$), while LyE of the 6–7-year-old group was also significantly lower than that of the 8–10-year-old group ($p = 0.032$). At the hip joint, post hoc comparisons showed that LyE of the 8–10-year-old group was significantly greater than that of both the 2–3-year-old group ($p = 0.006$) and the 4–5-year-old group ($p = 0.048$). At the knee joint, LyE was significantly lower for the 2–3-year-old group compared to that of both the 6–7-year-old group ($p = 0.043$) and the 8–10-year-old group ($p = 0.002$). There were no significant differences in LyE for the other group comparisons at the respective joints. There was a significant linear effect of age on LyE at the ankle (r = 0.374, $p = 0.05$), at the hip (r = 0.470, $p < 0.05$), and at the knee (r = 0.445, $p < 0.05$).

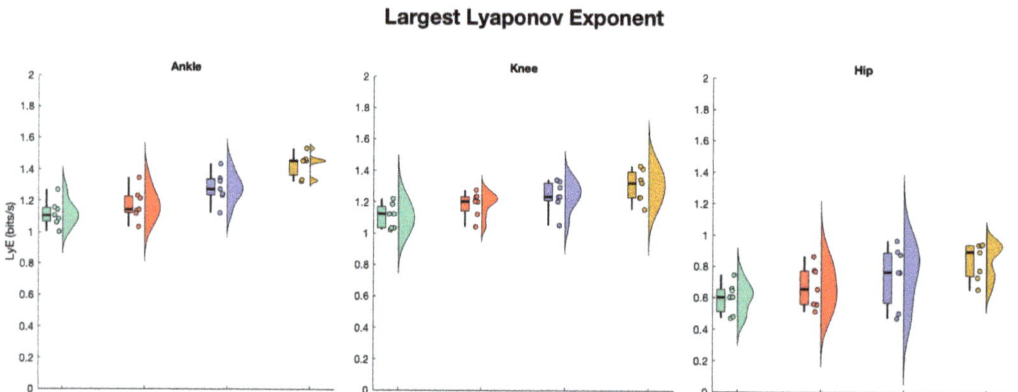

Figure 2. Violin and box plots showing the distribution of the largest Lyapunov exponent of the ankle, hip, and knee joint time series. Data are reported for the age groups.

Normalized comparisons for LyE showed that a significant effect remained for the ankle F (3,24) = 7.592, $p = 0.001$, and knee F (3,24) = 17.533, $p < 0.001$, joint angle time series.

Post hoc comparisons for the normalized LyE at the ankle revealed that the 2–3-year-old group exhibited a significantly lower normalized LyE than did the 6–7-year-old and 8–10-year-old groups ($p < 0.05$). The 4–5-year-old group also displayed a significantly lower normalized LyE at the ankle compared to that of the 6–7-year-old and 8–10-year-old groups ($p < 0.05$). There were no differences between the 2–3-year-old group and the 4–5-year-old group or the 6–7-year-old group and the 8–10-year-old group ($p > 0.05$) for the normalized LyE at the ankle. At the knee joint, post hoc comparisons showed that the normalized LyE significantly decreased with age. Specifically, the 2–3-year-old group showed a significantly greater normalized LyE compared to that of the 4–5-year-old, the 6–7-year-old, and the 8–10-year-old groups ($p < 0.05$). The 4–5-year-old group showed a significantly greater normalized LyE at the knee compared to that of both the 6–7-year-old and the 8–10-year-old groups ($p < 0.05$). There were no significant differences between the 2–3-year-old group and the 4–5-year-old group or the 6–7-year-old group and the 8–10-year-old group.

3.3. Recurrence Quantification Analysis of Joint Angle Time Series

The results for the RQA analysis are shown in Figure 3. There was a significant effect for %Det of the joint angle time series at the ankle $F(3,24) = 5.326$, $p = 0.006$. Specifically, post hoc tests showed that the 2–3-year-old group exhibited significantly less %Det than did both the 6–7-year-old group ($p = 0.040$) and the 8–10-year-old group ($p = 0.005$). No significant differences were found between the other age groups. There was also a significant effect for %Det at the hip joint $F(3,24) = 26.072$, $p < 0.001$. Specifically, post hoc tests show that the 2–3-year-old group showed a significantly lower %Det at the hip compared to both the 6–7-year-old group ($p < 0.001$) and the 8–10-year-old group ($p < 0.001$). The 4–5-year-old group also showed a significantly lower %Det than either the 6–7-year-old group ($p = 0.003$) or the 8–10-year-old group ($p < 0.001$). There were no other significant differences at the hip joint for %Det between groups. At the knee joint, there was not a significant effect for %Det. There was a significant linear trend for %Det of the ankle ($r = 0.628$, $p < 0.05$) and the hip ($r = 0.864$, $p < 0.05$), but not at the knee ($p > 0.05$).

Normalized comparisons for %Det showed there were significant effects for the ankle $F(3,24) = 14.076$, $p < 0.001$, hip $F(3,24) = 30.874$, $p < 0.001$, and knee $F(3,24) = 10.518$, $p < 0.001$ joints. Specifically, at the ankle joint, normalized %Det was significantly greater for the 2–3-year-old group than for the 6–7-year-old group and the 8–10-year-old group ($p < 0.05$). The 4–5-year-old group also showed significantly greater normalized %Det compared to the 6–7-year-old-group and the 8–10-year-old group ($p < 0.05$), but it was not different from that of the 2–3-year-old group. There were also no differences for normalized %Det at the ankle for the 6–7-year-old and 8–10-year-old groups ($p > 0.05$). At the hip joint, normalized %Det was significantly greater in the 2–3-year-old groups than in all three other groups ($p < 0.05$). The 4–5-year-old group also showed significantly greater normalized %Det than did the 6–7-year-old group and the 8–10-year-old group ($p < 0.05$), but there was no difference between the 6–7-year-old group and the 8–10-year-old group ($p > 0.05$). At the knee joint, normalize %Det was significantly greater in the 2–3-year-old group than in the 6–7-year-old group and the 8–10-year-old group ($p < 0.05$), but the 2–3-year-old group did not differ from from the 4–5-year-old group. The 4–5-year-old group showed significantly greater %Det at the knee joint than did the 6–7-year-old group and the 8–10-year-old group ($p < 0.05$), while the results for the 6–7-year-old group and the 8–10-year-old group did not differ ($p > 0.05$).

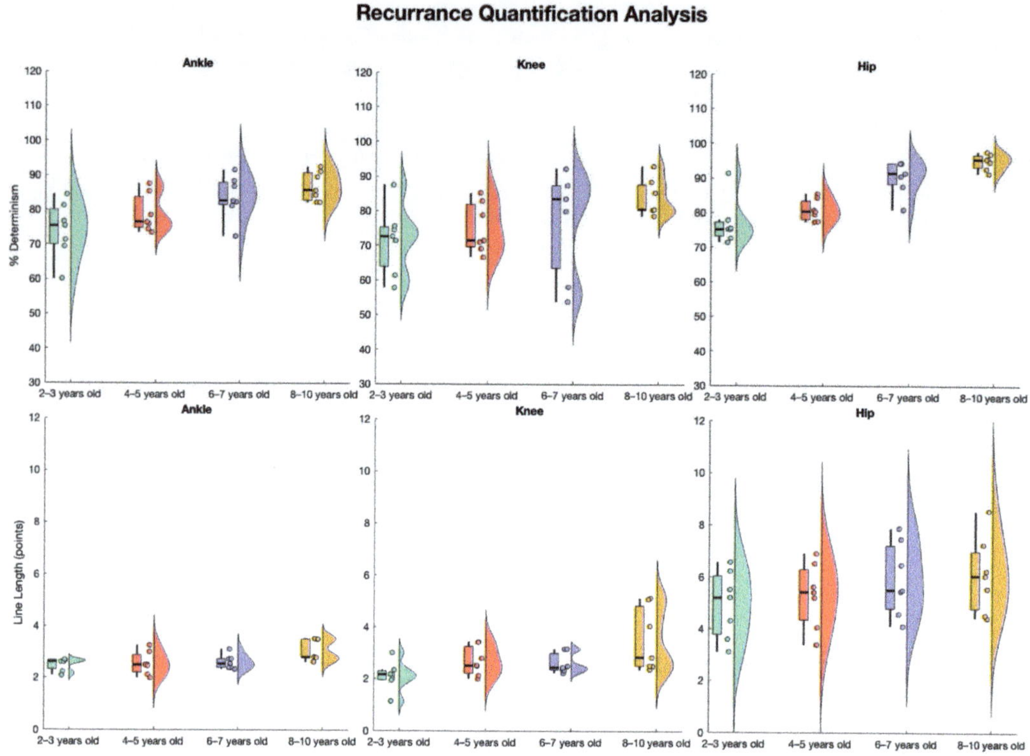

Figure 3. Violin and box plots showing the distribution of the recurrence quantification analysis (%Determinism, mean line) of the ankle, hip, and knee joint time series. Data are reported for the age groups.

There was a significant effect for MLine at the ankle joint F (3,24) = 3.773, $p = 0.024$. Specifically, post hoc tests showed that the 8–10-year-old group had a significantly greater MLine at the ankle joint compared to that of the 2–3-year-old group ($p = 0.031$). There were no other significant group differences for ankle MLine. There was not a significant effect for MLine at the hip joint ($p > 0.05$). There was a significant effect for MLine at the knee joint F (3,24) = 4.175, $p = 0.016$. Specifically, post hoc comparisons show that the 8–10-year-old group displayed a significantly greater MLine at the knee joint compared to that of the 2–3-year-old group ($p = 0.010$). No other significant differences existed between groups for MLine at the knee joint. There was a significant linear trend for MLine at the ankle (r = 0.489, $p < 0.05$) and the knee (r = 0.645, $p < 0.05$) but not at the hip joint ($p > 0.05$).

The normalized comparisons for MLine revealed a significant effect at the ankle joint F (3,24) = 11.826, $p < 0.001$. Post hoc comparisons of the normalized MLine at the ankle joint revealed that the 2–3-year-old group exhibited a significantly greater normalized MLine compared to that of the 4–5-year-old, 6–7-year-old, and 8–10-year-old groups ($p < 0.05$). The 4–5-year-olds also showed a significantly greater normalized MLine compared to that of the 6–7-year-old and the 8–10-year-old groups ($p < 0.05$). There was not a significant difference between the 6–7-year-old and the 8–10-year-old groups ($p > 0.05$).

4. Discussion

The purpose of our study was to assess the development of joint kinematic gait variability in typically developing children using nonlinear methods of analysis. We specifically wanted to investigate the joint angle time series of the lower extremities of children, at various developmental stages, while walking. Previous studies had investigated the spatiotemporal aspect of children's gait throughout development using nonlinear methods, but little emphasis has been placed on the joint angle time series of walking. We hypothesized that there would be an age effect on the gait variability of children. Specifically, as children aged, their gait variability would become more regular and more stable, as well as exhibit less stride-to-stride fluctuation.

Our hypotheses were partially supported for this study. Similar to results obtained by analysis of the spatiotemporal aspect of children's gait variability [8,10], the nonlinear analysis of the joint angle time series showed that the structure of gait variability in children is age-dependent. There was an age effect on the LyE and %Det of the joint angle time series at the ankle, hip, and knee. All three joints showed an increase with age, from youngest to oldest, for both the LyE and %Det. Neither the SE or MLine showed the same pattern of results or possessed the same significant differences. This is especially interesting because the SE of the stride time and stride length time series was age-dependent in the results of previous work [10]. Our results indicated a significant age effect on the nonlinear variability of the joint angle time series, even after normalizing for both leg length and gait speed to account for the natural differences in children due to growth and physical variations.

Interestingly, the direction of the age-dependency of the spatiotemporal variables of previous work contrasts with many of the results found in this study using the kinematic variables. The variability of the spatiotemporal variables decreased with age, while the variability of the kinematic variables increased as the children got older. These results could point to a hierarchy of behaviors to accomplish the desired goal of walking. To achieve the most stable gait, spatiotemporal variability may need to be minimized. Thus, as children age, the variability within their spatiotemporal gait decreases. How they accomplish this may be explained by the increase in their kinematic variability. Using the framework of dynamical systems theory [21], altering the parameter of kinematic variability, or in this case, increasing the complexity of the overall movement patterns, results in less variability in the spatiotemporal variables and an increased stability of gait.

As expected, our results indicate that gait variability and the structure of gait variability are continually changing throughout development in children and are extremely age-dependent. The LyE is able to examine the quality and structure of movement patterns and movement stability [12] and in this case, the joint angles of gait in children. Larger magnitudes of LyE indicate greater attractor divergence of the gait patterns and can be equated with maturing control of the motor system. As our groups increased in age from 2 to 10 years old, the LyE also increased. This was evident at the ankle, hip, and knee joints, respectively. As children gain more walking experience, the quality of their movements become more refined, exhibiting more stable, yet complex, movements. As children age, their gait variability may be becoming more chaotic, but it is also becoming more deterministic as well.

In our study, the RQA measure of %Det proved to also increase as children aged. Like the LyE outcomes, the %Det results span across all three joints of the lower extremity. %Det can be interpreted as a decrease in variability as a system becomes more deterministic. This seems to contradict and go against the premise of this study. However, when coupled with the LyE results, the overall results of our study agree with the theory of optimal movement variability. The theory of optimal movement variability posits that there is a sweet spot of sorts for movement variability. Too much or too little variability is unfavorable and detrimental to the system, as evidenced by a connection with unhealthy systems. Interpreting our results using this theoretical basis shows that the time series is increasingly diverging, while also becoming more deterministic. This behavior shows the complexity and sophistication of the development of movement trajectories.

As children get older and gain more experience walking, their neuromuscular systems and overall motor control are also maturing [22]. The combination of the maturation of the systems and gained experience leads to an overall better organization of movements [9,10]. The increase in divergent patterns, as well as increased determinism, makes for a more robust movement system that is capable of dealing with small perturbations without flaw. When we get older and become more experienced walkers, we can navigate our environment with ease. Small increases in rise or bumps in a path that could elicit falls in the youngest of walkers are hardly noticed by the more adept, experienced walker.

Our results provide a blueprint for the developmental trajectory of gait variability in typically developing children. Investigations into the maturation of gait and the development of movement pathology in children can be weighed against our results to better understand how pathology affects gait variability. This information can then be used to further understand the mechanisms underlying that pathology and aid and assist with the creation of therapies and new movement strategies to eventually overcome pathology. Future studies should consider exploration into the development of neuromuscular control utilizing the combination of nonlinear methods of analysis of children's gait and other measures, such as electromyography. A multifaceted approach to researching the development of motor control in children, utilizing nonlinear analysis, could shed light on many of the unknowns regarding how children self-organize and how their motor movements develop organically throughout childhood. Limitations to our study include the children's potential lack of experience walking on the treadmill. Although all four groups were equally inexperienced walking on a treadmill, to minimize this effect, all participants were given an acclimation period of walking until comfort was perceived. Another limitation is that the children were provided with the specific footwear used for this study to control for varying types of shoes that the children currently wore. The effect of the new shoe was minimized through the acclimation period on the treadmill.

5. Conclusions

In conclusion, we set out to investigate the development of gait variability in children by analyzing the stride-to-stride dynamics of the joint angle times series. Our study was able to advance the understanding of the developmental trajectory of children's gait variability by utilizing nonlinear methods of analysis. Children's gait becomes more refined with age, gaining sophistication by increasing adaptability, as well as organization. Walking experience alone is not the driving force, as many systems within the developing child are maturing. This type of investigation merely scratches the surface regarding obtaining a full understanding of the maturation dynamics of children's gait. A multifaceted approach to understanding motor control should be utilized in future research.

Author Contributions: Conceptualization, B.C.A. and A.K.; methodology, B.C.A. and A.K.; software, B.C.A. and Z.L.M.; validation, B.C.A., Z.L.M. and A.K.; formal analysis, B.C.A. and A.K.; investigation, B.C.A., Z.L.M. and A.K.; resources, B.C.A.; data curation, B.C.A. and Z.L.M.; writing—original draft preparation, B.C.A.; writing—review and editing, B.C.A., Z.L.M. and A.K.; visualization, B.C.A.; supervision, B.C.A. and A.K. All authors have read and agreed to the published version of the manuscript.

Funding: This research received no external funding.

Institutional Review Board Statement: The study was conducted in accordance with the Declaration of Helsinki and approved by the Institutional Review of The University of Nebraska Medical Center (IRB #682-14-EP).

Informed Consent Statement: Parental consent and child consent or assent was obtained from all subjects involved in the study.

Data Availability Statement: Data are available upon request due to privacy restrictions.

Conflicts of Interest: The authors declare no conflicts of interest.

References

1. Stergiou, N.; Harbourne, R.; Cavanaugh, J. Optimal movement variability: A new theoretical perspective for neurologic physical therapy. *J. Neurol. Phys. Ther.* **2006**, *30*, 120–129. [CrossRef] [PubMed]
2. Stergiou, N.; Kent, J.A.; McGrath, D. Human movement variability and aging. *Kinesiol. Rev.* **2016**, *5*, 15–22. [CrossRef]
3. Goldberger, A.L.; Amaral, L.A.N.; Hausdorff, J.M.; Ivanov, P.C.; Peng, C.-K.; Stanley, H.E. Fractal dynamics in physiology: Alterations with disease and aging. *Proc. Natl. Acad. Sci. USA* **2002**, *99* (Suppl. S1), 2466–2472. [CrossRef] [PubMed]
4. Peng, C.-K.; Havlin, S.; Stanley, H.E.; Goldberger, A.L. Quantification of scaling exponents and crossover phenomena in nonstationary heartbeat time series. *Chaos* **1995**, *5*, 82–87. [CrossRef] [PubMed]
5. Kaipust, J.P.; Huisinga, J.M.; Filipi, M.; Stergiou, N. Gait variability measures reveal differences between multiple sclerosis patients and healthy controls. *Mot. Control* **2012**, *16*, 229–244. [CrossRef] [PubMed]
6. Kaipust, J.P.; McGrath, D.; Mukherjee, M.; Stergiou, N. Gait variability is altered in older adults when listening to auditory stimuli with differing temporal structures. *Ann. Biomed. Eng.* **2013**, *41*, 1595–1603. [CrossRef] [PubMed]
7. Wurdeman, S.R.; Huben, N.B.; Stergiou, N. Variability of gait is dependent on direction of progression: Implications for active control. *J. Biomech.* **2012**, *45*, 653–659. [CrossRef] [PubMed]
8. Wurdeman, S.R.; Myers, S.A.; Jacobsen, A.L.; Stergiou, N. Prosthesis preference is related to stride-to-stride fluctuations at the prosthetic ankle. *J. Rehabil. Res. Dev.* **2013**, *50*, 671–686. [CrossRef] [PubMed]
9. Hausdorff, J.M.; Zemany, L.; Peng, C.-K.; Goldberger, A.L. Maturation of gait dynamics: Stride-to-stride variability and its temporal organization in children. *J. Appl. Physiol.* **1999**, *86*, 1040–1047. [CrossRef] [PubMed]
10. Sutherland, D.H.; Olshen, R.; Biden, E.; Wyatt, M. *The Development of Mature Walking*; Cambridge University Press: Cambridge, UK, 1988.
11. Applequist, B.C.; Motz, Z.L.; Kyvelidou, A. Spatiotemporal Gait Variability in Children Aged 2 to 10 Decreases throughout Pre-Adolescence. *Biomechanics* **2023**, *3*, 571–582. [CrossRef]
12. Myers, S.A.; Stergiou, N.; Pipinos, I.I.; Johanning, J.M. Gait variability patterns are altered in healthy young individuals during the acute reperfusion phase of ischemia-reperfusion. *J. Surg. Res.* **2010**, *164*, 6–12. [CrossRef] [PubMed]
13. Buzzi, U.H.; Stergiou, N.; Kurz, M.J.; Hageman, P.A.; Heidel, J. Nonlinear dynamics indicates aging affects variability during gait. *Clin. Biomech.* **2003**, *18*, 435–443. [CrossRef] [PubMed]
14. Wurdeman, S.R.; Myers, S.A.; Jacobsen, A.L.; Stergiou, N. Adaptation and prosthesis effects on stride-to-stride fluctuations in amputee gait. *PLoS ONE* **2014**, *9*, e100125. [CrossRef] [PubMed]
15. Dingwell, J.B.; Cusumano, J.P.; Cavanagh, P.R.; Sternad, D. Local dynamic stability versus kinematic variability of continuous overground and treadmill walking. *J. Biomech. Eng.* **2001**, *123*, 27–32. [CrossRef] [PubMed]
16. Hein, T.; Grau, S. Can minimal running shoes imitate barefoot heel-toe running patterns? A comparison of lower leg kinematics. *J. Sport Health Sci.* **2014**, *20*, 67–73. [CrossRef]
17. Hollman, J.H.; Watkins, M.K.; Imhoff, A.C.; Braun, C.E.; Akervik, K.A.; Ness, D.K. A comparison of variability in spatiotemporal gait parameters between treadmill and overground walking conditions. *Gait Posture* **2016**, *43*, 204–209. [CrossRef] [PubMed]
18. Nigg, B.M.; Cole, G.K.; Nachbauer, W. Effects of arch height of the foot on angular motion of the lower extremities in running. *J. Biomech.* **1993**, *26*, 909–916. [CrossRef] [PubMed]
19. Vaughan, C.L.; Davis, B.L.; O'Connor, J.C. *Dynamics of Human Gait*; Human Kinetics: Champaign, IL, USA, 1992.
20. Rapp, P.E.; Albano, A.M.; Schmah, T.I.; Farwell, L.A. Filtered noise can mimic low-dimensional chaotic attractors. *Phys. Rev. E* **1993**, *47*, 2289–2297. [CrossRef] [PubMed]
21. Stergiou, N.; Buzzi, U.H.; Kurz, M.J.; Heidel, J. Nonlinear tools in human movement. In *Innovative Analyses of Human Movement*; Human Kinetics: Champaign, IL, USA, 2003; pp. 63–90.
22. Thelen, E. Motor development: A new synthesis. *Am. Psychol.* **1995**, *50*, 79–95. [CrossRef] [PubMed]

biomechanics

MDPI

Article

Asymmetric Knee Joint Loading in Post-Stroke Gait: A Musculoskeletal Modeling Analysis of Medial and Lateral Compartment Forces

Georgios Giarmatzis [1,*], Nikolaos Aggelousis [1], Marinos Marinidis [1], Styliani Fotiadou [2], Erasmia Giannakou [1], Evangelia Makri [1], Junshi Liu [3] and Konstantinos Vadikolias [2]

1 Department of Physical Education and Sport Science, Democritus University of Thrace, 69100 Komotini, Thrace, Greece; nagelous@phyed.duth.gr (N.A.); mmarinid@phyed.duth.gr (M.M.); egiannak@phyed.duth.gr (E.G.); emakri@phyed.duth.gr (E.M.)
2 Department of Neurology, School of Medicine, Democritus University of Thrace, 68100 Alexandroupolis, Thrace, Greece; sfotiado@phyed.duth.gr (S.F.); kvadikol@med.duth.gr (K.V.)
3 Department of Physical Education, Dongguan University of Technology, Dongguan 523808, China; junshi100610@gmail.com
* Correspondence: ggiarmat@phyed.duth.gr

Abstract: Background/Objectives: Stroke survivors often develop asymmetric gait patterns that may lead to abnormal knee joint loading and potentially increased risk of osteoarthritis. This study aimed to investigate differences in knee joint loading between paretic and non-paretic limbs during walking in individuals post-stroke. **Methods**: Twenty-one chronic stroke survivors underwent three-dimensional gait analysis. A modified musculoskeletal model with a specialized knee mechanism was used to estimate medial and lateral tibiofemoral contact forces during the stance phase. Statistical parametric mapping was used to identify significant differences in joint kinematics, kinetics, and contact forces between limbs. Stepwise regression analyses examined relationships between knee moments and compartmental contact forces. **Results**: Significant differences in knee loading were observed between limbs, with the non-paretic limb experiencing higher medial compartment forces during early stance (6.7–15.1%, $p = 0.001$; 21.9–30.7%, $p = 0.001$) and late stance (72.3–93.7%, $p < 0.001$), and higher lateral compartment forces were recorded during pre-swing (86.2–99.0%, $p < 0.001$). In the non-paretic limb, knee extensor moment was the primary predictor of first peak medial contact force ($R^2 = 0.573$), while knee abductor moment was the primary predictor in the paretic limb ($R^2 = 0.559$). **Conclusions**: Musculoskeletal modeling revealed distinct asymmetries in knee joint loading between paretic and non-paretic limbs post-stroke, with the non-paretic limb experiencing consistently higher loads, particularly during late stance. These findings suggest that rehabilitation strategies should address not only paretic limb function but also potentially harmful compensatory mechanisms in the non-paretic limb to prevent long-term joint degeneration.

Keywords: stroke rehabilitation; knee joint loading; musculoskeletal modeling; gait biomechanics

1. Introduction

Stroke remains one of the leading causes of long-term disability worldwide, resulting in significant impairments in sensory, motor, cognitive, and visual functions that substantially impact daily activities [1,2]. A particularly challenging consequence of stroke is asymmetric gait patterns, where survivors predominantly rely on their unaffected limb for locomotion, leading to what is known as hemiplegic gait [3]. While, this compensatory

https://doi.org/10.3390/biomechanics5020039

strategy enables mobility and allow patients to retain steady walking state, it may introduce biomechanical alterations in both lower limbs, such as abnormal knee loading. Despite achievements in rehabilitation techniques [4,5], stroke survivors still encounter significant abnormalities in their walking patterns, such as altered joint mobility, spatiotemporal asymmetries, uneven weight distribution on the lower limbs, and modifications in muscle activation patterns [6].

Abnormal walking pattern in unilateral conditions and the repetitive high loads imposed on joint structures can lead to stress-related joint injuries and disorders in the unaffected limb or disease progression in the affected limb. A similar pattern has been observed in individuals with unilateral osteoarthritis (OA) or lower-limb amputation, who are at higher risk of developing OA in the contralateral knee joint (i.e., the knee of the opposite limb from the affected side). This supports the belief that asymmetric mechanical loading plays an important role in the development of knee OA [7–9]. However, these findings are based on external measures as proxies for actual joint loading, such as external knee joint moments [10]. Extensive research on knee loading during gait of knee OA patients has revealed knee adduction and flexion moments (KAM and KFM, respectively) which only relate to first peak of knee contact forces [11], whereas no statistical relationship has been found between the second knee contact force (KCF) peak and external joint moments. Moreover, a case study using instrumented knee prosthesis showed that a reduction in first peak KAM does not guarantee a reduction in medial contact load [12] or indicate changes in the relative distribution of the loading between medial and lateral compartments. Lastly, some studies in patients with early stages of knee OA suggest that altered KAM and KFM are not risk factors in the initial development of knee OA [13].

The characteristic impairments of post-stroke gait, including reduced walking speed, altered knee joint range of motion [14], abnormal muscle co-activation patterns [15], and asymmetrical knee joint kinetics between paretic and non-paretic limbs [16], could increase joint loading and contribute to cartilage degeneration [17]. Knee cartilage thinning has been reported in hemiparetic knees [18,19], particularly on the lateral side [20], and is linked to knee pain and knee OA onset or progression. Knee OA prevalence among stroke patients can reach up to 21% [21], which can negatively affect rehabilitation outcome [22] and quality of life. A previous study [23] in stroke survivors showed that there is likely to be an increase in loading either in the non-paretic limb or in the paretic limb during gait compared to healthy individuals, while in another study [24], no different knee loading was found between lower limbs. Again, these studies rely on knee external joint moments, which correlate only with the first peak of the medial KCF [25], leaving other knee loading parameters—such as complete waveforms and peak lateral KCF—unexplored. Furthermore, the relationship between external moments and peak knee loading in stroke gait remains unclear.

Musculoskeletal (MSK) modeling offers a more sophisticated approach to estimating internal joint forces, accounting for both external loads and muscle contributions, thus providing more accurate insights into the actual cartilage loading conditions experienced during gait [26]. It is primarily based on accurate recorded motion from 3D gait analysis labs using retroreflective markers and devices recording ground reaction forces, as inputs to inverse dynamics algorithms that calculate muscle and joint reaction forces utilizing a wealth of MSK models [27,28]. However, despite its potential and its rather limited use in examining post-stroke gait [29], there is currently no research applying musculoskeletal modeling to examine knee joint loading in stroke survivors. Nevertheless, it has been extensively used to calculate knee contact forces in numerous scenarios [30,31], in particular for knee OA and total knee replacement patients, and validated against actual forces recorded from a limited number of total joint replacement devices [32].

Therefore, the purpose of this study was to investigate the differences in knee joint loading between paretic and non-paretic limbs during walking in individuals post-stroke using musculoskeletal modeling. We hypothesized that the non-paretic limb would experience greater knee joint forces compared to the paretic limb, reflecting the compensatory strategies commonly adopted by stroke survivors during gait. As a secondary aim, we explored the relationship between external joint moments and knee joint loading and tried to validate the findings of the previous literature on knee OA patients and healthy individuals. These insights could inform training protocols to optimize stroke recovery.

2. Materials and Methods

2.1. Participants

Twenty-one individuals (eleven males) with chronic stroke participated in this study. Participants were recruited from the outpatient Neurological Rehabilitation Unit at the University Hospital of Alexandroupolis, Greece and participated in two separate investigations, with eleven participants from a previous repeatability study [33] and ten from a subsequent investigation. The participants' affected side was distributed as follows: eleven had left hemiparesis and ten had right hemiparesis. The mean age was 62.8 ± 4.7 years and the mean body mass index was 28.9 ± 4.14. All participants provided written informed consent, and the study protocols were approved by the institutional ethics committee of the Democritus University of Thrace, Greece. Participants were included if they met the following criteria: (1) chronic phase of stroke (at least 6 months post-stroke), (2) age above 18 years, (3) walking speed above 0.2 m/s with no upper limit, (4) ability to walk without assistance, and (5) diagnosed hemiparesis with observable motor impairment in the affected limb at the time of enrolment, as confirmed by a licensed healthcare professional (6). Participants with only non-motor stroke symptoms (such as isolated facial weakness, dysarthria, or sensory deficits) were not eligible for the study. According to the National Institutes of Health Stroke Scale [32], almost all patients had a score between 1 and 4 (minor stroke) with one patient scoring 7 (moderate stroke).

2.2. Data Collection

Three-dimensional gait analysis was performed using two different motion capture setups. For the first cohort, three-dimensional marker coordinates were recorded using six infrared cameras (Vicon MX 0306012, Oxford, UK) sampling at 100 Hz. For the second cohort, a 10-camera Vicon (Vicon Motion Systems Ltd., Oxford, UK) system was employed, also sampling at 100 Hz. In both setups, ground reaction forces were recorded using two force plates (type 9281B11 and 9281CA, Kistler Instruments AG, Winterthur, Switzerland) embedded in the middle of a 10-m walkway, sampling at 1000 Hz.

Retroreflective markers were placed on anatomical landmarks following two protocols (see Figure 1). For the first cohort, markers were placed according to the Vicon Plug-in-Gait lower-body protocol, while the second cohort utilized a full-body marker according to the conventional gait model set comprising 57 markers, although only the markers corresponding to the PiG model were used for the analysis. All participants walked barefoot along the walkway at their self-selected speed. Marker trajectories and ground reaction forces were processed using Vicon Nexus software (v 2.12.1) and filtered using a low-pass filter with a 6 Hz cutoff frequency. A minimum of five successful trials were collected for each participant, where a successful trial was defined as clean force plate contact with the affected limb.

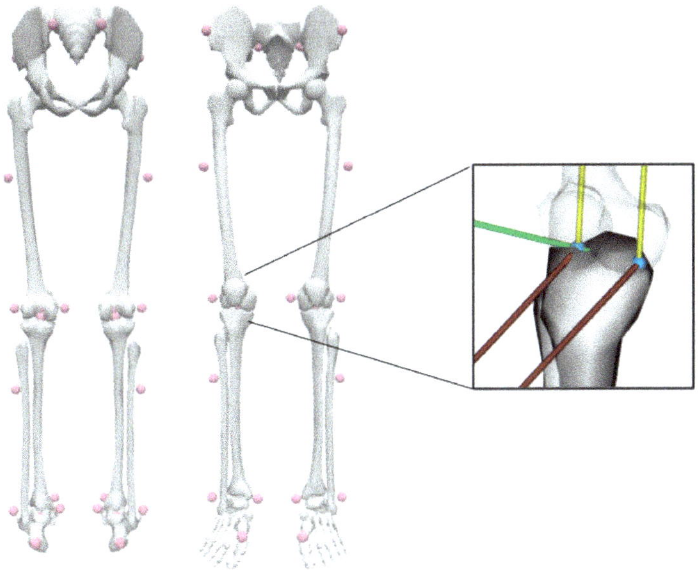

Figure 1. Left: Full-body marker placement according to the Plug-in-Gait protocol with retroreflective markers (pink spheres) positioned on anatomical landmarks. Right: Detailed view of the specialized knee mechanism used to estimate medial and lateral compartment contact forces. The mechanism features medial and lateral contact points (blue spheres) and the frontal plane alignment components that enable distribution of forces between compartments.

2.3. Musculoskeletal Modeling

A modified version of the Lerner model [34] was used to estimate joint kinematics, kinetics, and knee contact forces. The original model was adapted by condensing the torso segments to the pelvis due to marker protocol restrictions in the lower limb. The model included a ball-and-socket joint between the third and fourth lumbar vertebra, three translations and three rotations of the pelvis, ball-and-socket joints at the hips, and hinge ankle and subtalar joints.

The knee mechanism was specifically designed to resolve medial and lateral tibiofemoral contact forces. While sagittal plane rotation and translations of the tibia and patella relative to the femur followed Delp's specifications [35], the tibiofemoral mechanism was augmented with a distal femoral component and a tibial plateau body. These components enabled configuration of frontal plane alignment through orientation parameters in both the femur and tibia. The tibiofemoral articulation was modeled using a series of joints, namely a primary knee joint, defining sagittal plane rotations and translations, and two hinge joints connecting the sagittal articulation frame to medial and lateral tibiofemoral compartments. These hinge joints, with axes perpendicular to the frontal plane, were welded at the anterioposterior mid-point of the tibial plateaus, maintaining fixed positions relative to the tibia while articulating with the femoral component during flexion–extension. While individual hinge joints could not resist frontal plane moments, their parallel arrangement allowed load sharing between medial and lateral compartments via medial and lateral contact points (see Figure 1) being positioned at the anterioposterior mid-point of the respective tibial plateaus, resolving the net reaction forces and frontal plane moments across the tibiofemoral joint. The knee remained a single degree-of-freedom joint with motion restricted to the sagittal plane.

Marker trajectories and ground reaction forces were used as input in OpenSim 3.3. The model was scaled to each participant's anthropometry using markers from a recorded static trial prior to recordings. For each successful walking trial, joint angles were calculated using inverse kinematics, solving for the generalized coordinates that minimized the weighted sum of squared differences between experimental and model marker positions, according to the following equation:

$$\min_{q}\left[\sum_{i\in markers} w_i \left\| x_i^{exp} - x_i(q)^2 \right\| + \sum_{j\in unprescribed\ coords} \omega_j \left(q_j^{exp} - q_j\right)^2\right] \tag{1}$$

$$q_j = q_j^{exp}\ for\ all\ prescribed\ coordinates\ j$$

where q is the vector of generalized coordinates being solved for, x_i^{exp} is the experimental position of marker i, $x_i(q)$ is the position of the corresponding model marker (which depends on the coordinate values), and q_j^{exp} is the experimental value for coordinate j. Prescribed coordinates are set to their experimental values. Joint moments were computed using the inverse dynamics tool which solves the classical equations of motion, as follows:

$$M(q)\ddot{q}+C(q,\dot{q})+G(q) = \tau \tag{2}$$

where M(q) is the system mass matrix, $C(q, \dot{q})$ is the vector of the Coriolis and centrifugal forces, G(q) is the vector of gravitational forces, and τ is the vector of generalized forces (joint moments). Given the known motion $(q, \dot{q}, \ddot{q})$, the tool solves for the unknown generalized forces. Then, static optimization was performed to estimate muscle forces, using a cost function that minimized the sum of squared muscle activations, according to the following Equation (3):

$$\min_{a} J(a) = \sum_{i=1}^{n} (a_i)^2, \tag{3}$$

where

- a_i is the activation level of muscle i.
- n is the total number of muscles in the model.

The model is subject to the following constraints:

1. Moment equilibrium:

$$\sum_{i=1}^{n} r(s)_i \times F_i(s,a,i) = M \tag{4}$$

where

- $r_i(s)$ is the moment arm vector of muscle i at joint configuration s.
- $F_i(s,a_i) = a_i \cdot F_{max,i} \cdot f_l(l_i) \cdot f_v(v_i)$ is the force produced by muscle i.
- $F_{max,i}$ is the maximum isometric force of muscle i.
- $f_l(l_i)$ is the force–length relationship.
- $f_v(v_i)$ is the force–velocity relationship.
- M is the net joint moment derived from inverse dynamics.

2. Activation bounds: $0 \leq a_i \leq 1$ for all $i \in \{1, 2, \ldots, n\}$

Finally, Opensim's joint reaction analysis was applied to each compartment's hinge joint to calculate medial and lateral contact forces, normalized to body mass. The parallel arrangement of the two hinge joints allows them to share all loads transmitted between the femur and tibia, automatically resolving the medial and lateral contact forces required to balance the net reaction forces and frontal plane moments across the tibiofemoral joint.

2.4. Statistical Analysis

Joint kinematics, kinetics, and vertical knee contact forces were time-normalized to 100% of the stance phase. Group differences between paretic and non-paretic limbs were assessed using statistical parametric mapping (SPM) [36] and a paired t-test via the open-source spm1d code [37], enabling statistical analysis and comparison of the waveform data across the entire stance phase by conducting point-by-point hypothesis testing while controlling for multiple comparisons [36].

For kinematic analysis, we examined hip flexion–extension, hip abduction-adduction, hip internal-external rotation, knee flexion–extension, and ankle dorsi–plantarflexion angles. Kinetic variables included the corresponding joint moments, with the addition of the knee abductor moment. For knee contact forces, both medial and lateral compartment forces were analyzed.

Statistical significance was set at $\alpha = 0.05$. The SPM analysis generated statistical parametric maps (SPM{t}) to identify regions of significant differences between limbs. Significance clusters were defined as temporal regions where the SPM{t} statistic exceeded the critical threshold. The temporal location, duration, critical t-values (t^*), and p-value are reported for each significant cluster.

Relationships between knee contact forces and joint moments were examined through a stepwise regression analysis. For each trial, two peaks were identified in both medial and lateral compartment forces: the first peak was identified during early stance (5–50%) and the second peak was identified during late stance (51–100%). At each peak time point, the corresponding knee extensor and knee abductor moments were extracted. The stepwise regression first identified the strongest predictor (either knee extensor or abductor moment) based on adjusted R^2 values. The second predictor was then added to assess the incremental contribution to the model fit. For each model, we calculated the initial R^2, the R^2 change (ΔR^2) after adding the second predictor, and the statistical significance of the R^2 change through an F-test. The latter was calculated as $((R^2 \text{ change})/(1))/((1 - R^2\text{full})/(n-3))$, where R^2 change is the improvement in R^2 after adding the second predictor, R^2full is the R^2 of the complete model, and n is the sample size. This analysis was performed separately for paretic and non-paretic limbs. All contact forces were normalized to body weight, and moments were expressed in Nm/kg. Statistical significance was set at $\alpha = 0.05$. All statistical analyses were performed using custom Python (v3.8) scripts using the open-source spm1d package (v 0.4) [36] and statsmodels packages [38].

Given that our primary analysis compared paretic versus non-paretic limbs within the same individuals, the side of hemiparesis was inherently controlled for in the study design. The narrow age distribution of our cohort (62.8 $\pm$ 4.7 years) minimized potential age-related confounding effects. Joint moments and contact forces were appropriately normalized to body mass and weight, respectively, to control for inter-subject variability in anthropometric characteristics.

3. Results

Significant differences in knee joint contact forces were observed between paretic and non-paretic limbs during the stance phase. In the medial compartment, distinct periods of significant differences occurred during early and late stance. The lateral compartment exhibited differences only during late stance. Overall, medial compartment forces were substantially higher than lateral compartment forces in both limbs (Table 1).

Table 1. Peak knee contact forces and corresponding joint moments during stance phase. Values are presented as mean ± standard deviation for both medial and lateral compartments of paretic and non-paretic limbs. Contact forces are normalized to body weight (BW).

Variable	Paretic Limb	Non-Paretic Limb
Medial Compartment		
First peak contact force (BW)	2.37 ± 0.51	2.42 ± 0.43
Knee extensor moment at first peak (Nm/kg)	0.57 ± 0.34	0.63 ± 0.34
Knee abductor moment at first peak (Nm/kg)	0.42 ± 0.11	0.44 ± 0.11
Second peak contact force (BW)	2.58 ± 0.60	2.72 ± 0.47
Knee extensor moment at second peak (Nm/kg)	0.30 ± 0.22	0.33 ± 0.23
Knee abductor moment at second peak (Nm/kg)	0.34 ± 0.10	0.32 ± 0.10
Lateral Compartment		
First peak contact force (BW)	0.97 ± 0.41	0.88 ± 0.29
Knee extensor moment at first peak (Nm/kg)	0.39 ± 0.36	0.42 ± 0.33
Knee abductor moment at first peak (Nm/kg)	0.26 ± 0.28	0.21 ± 0.17
Second peak contact force (BW)	0.91 ± 0.41	1.03 ± 0.40
Knee extensor moment at second peak (Nm/kg)	0.41 ± 0.24	0.48 ± 0.26
Knee abductor moment at second peak (Nm/kg)	0.17 ± 0.13	0.20 ± 0.10

Statistical parametric mapping analysis revealed significant between-limb differences in knee contact forces during specific stance phases (see Figure 2). The medial compartment forces exhibited two distinct periods of significant differences. During early stance, differences were observed from 6.7% to 15.1% (t^*= 3.2, $p < 0.001$) and from 21.9% to 30.7% (t^*= 3.2, $p < 0.001$). The most pronounced difference in medial compartment forces occurred during terminal stance and pre-swing, spanning from 72.3% to 93.7% of stance (t^*= 3.2, $p < 0.001$). For the lateral compartment, a single period of significant difference was identified during pre-swing, extending from 86.2% to 99.0% of stance (t^*= 3.17, $p < 0.001$).

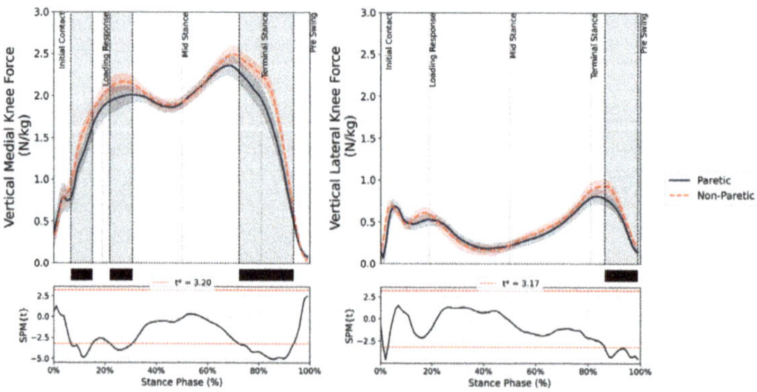

Figure 2. Knee joint contact forces throughout stance phase for paretic (solid blue line) and non-paretic (dashed red line) limbs. Left panel shows medial compartment forces and right panel shows lateral compartment forces. Forces are normalized to body mass (N/kg). Shaded areas represent 95% confidence intervals. Gray-shaded regions indicate periods of significant differences between limbs ($p < 0.05$) as determined by SPM. Vertical dotted lines denote stance phase events: initial contact, loading response, mid-stance, terminal stance, and pre-swing. Black bars along the x-axis highlight the temporal regions of significant differences. Lower panels display SPM{t} statistic curves (black lines) with critical threshold lines (red dashed) and values (t*).

3.1. Kinematics and Kinetics

Joint kinematics analysis revealed consistent asymmetrical patterns between limbs throughout stance. At the hip, flexion angle showed two distinct periods of differences, namely during early stance (0–36.9%, $t^* = 2.71$, $p = 0.002$) and late stance (59.2–91.2%, $t^* = 2.71$, $p = 0.004$), with higher angles on the non-paretic side. Higher non-paretic knee flexion angles were found during mid-stance (10.6–56.7%, $t^* = 2.72$, $p < 0.001$) and terminal stance (84.4–99%, $t^* = 2.72$, $p = 0.029$). Non-paretic ankle plantarflexion was significantly larger at initial contact (0–5.8%, $t^* = 2.72$, $p = 0.046$) and ipsilateral dorsiflexion during pre-swing (90.1–99%, $p = 0.042$).

Joint kinetics analysis revealed distinct bilateral asymmetries across all joints, being higher mainly on the non-paretic side. Joint moments are named according to the counteracting muscle group around the respective joint. Hip extensor moment was higher on the non-paretic side during loading response (11.7–19.8%, $p < 0.001$), and non-paretic hip abductor moment exhibited higher values during early stance (25.1–38.6%, $p < 0.001$) and pre-swing (81.9–90.6%, $p < 0.001$). Hip internal rotator moment demonstrated sustained differences through early and mid-stance (13–52%, $p < 0.001$), with larger values on the non-paretic side. Non-paretic knee extensor moment was higher during mid-stance (29.7–39.4%, $p < 0.001$) and terminal stance (86.5–96.3%, $p < 0.001$), while non-paretic knee abductor moment showed significantly larger values during early stance (24.5–38.3%, $p < 0.001$) and pre-swing (84–91.2%, $p = 0.002$). The ankle plantarflexor moment exhibited differences during loading response (8.6–15%, $p = 0.012$) and a prolonged period of bilateral asymmetry during terminal stance and pre-swing (65.2–95.3%, $p < 0.001$), with higher values on the non-paretic side.

Regression Analysis

The relationships between knee moments and contact forces showed distinct patterns between limbs, as shown in the regression plots in Figure A1 in Appendix A. In the paretic limb, the first peak medial contact force was predominantly explained by the knee abductor moment (R^2adj = 0.559, $p < 0.001$), with knee extensor moment providing additional explanatory power (ΔR^2adj = 0.079, $p < 0.001$), yielding a final R^2adj of 0.638. The second peak medial contact force demonstrated a weaker association with the knee abductor moment (R^2adj = 0.167, $p < 0.001$), with no significant contribution from the knee extensor moment (ΔR^2adj = 0.003, $p = 0.516$).

Lateral compartment forces in the paretic limb showed that the first peak was primarily associated with the knee abductor moment (R^2adj = 0.142, $p < 0.001$), with knee extensor moment contributing modestly (ΔR^2adj = 0.039, $p = 0.028$). The second peak exhibited an inverse pattern, with knee extensor moment as the primary predictor (R^2adj = 0.166, $p < 0.001$) and knee abductor moment adding minor explanatory power (ΔR^2adj = 0.035, $p = 0.036$). However, all associations between lateral contact forces and joint moments remained weak.

In the non-paretic limb, the first peak medial contact force was primarily associated with the knee extensor moment (R^2adj = 0.573, $p < 0.001$), further strengthened by the addition of knee abductor moment (ΔR^2adj = 0.099, $p < 0.001$), achieving a final R^2adj of 0.671. The second peak showed weaker associations, with knee abductor moment as the primary predictor (R^2adj = 0.109, $p < 0.001$) and knee extensor moment providing minimal additional explanation (ΔR^2adj = 0.041, $p = 0.027$).

Lateral compartment forces in the non-paretic limb were predominantly associated with knee extensor moments. The first peak showed a moderate relationship with knee extensor moment (R^2adj = 0.228, $p < 0.001$), with no significant contribution from knee abductor moment (ΔR^2adj = −0.005, $p = 1.000$). Similarly, the second peak was primarily explained by knee extensor moment (R^2adj = 0.227, $p < 0.001$), with knee abductor moment showing a non-significant contribution (ΔR^2adj = 0.020, $p = 0.099$).

4. Discussion

To our knowledge, this is the first study reporting compartmental knee joint forces during post-stroke gait. Our findings reveal distinct asymmetries in knee joint loading between paretic and non-paretic limbs, with the latter experiencing consistently higher loads, particularly during late stance. The increased loading in the non-paretic limb appears to be driven by compensatory mechanisms, as evidenced by the significant kinematic and kinetic differences observed between limbs.

Our musculoskeletal modeling analysis reveals that actual joint contact forces show distinct asymmetries between legs during multiple phases of stance. The non-paretic medial compartment exhibited significant higher loading during early and late stance than the paretic side, along with the corresponding lateral compartment which also showed higher loading during terminal stance. Higher non-paretic joint moments (see Figure 3) seen in early-to-mid and late stance phases may explain this finding, since internal muscle moments must be produced to counteract the external ones for each motion frame. From the motion strategy perspective, higher non-paretic hip flexion/extension and knee flexion angles during most parts of the stance phase could explain the higher moments on the sagittal plane due to the increased lever arm of the GRF, hence the elevated muscle forces, especially from knee extensors/hip flexors. Liu et al. [39] found that in particular vastii and rectus femoris muscles contributed the most to the medial joint forces of knee OA patients during walking.

The finding of increased loading in the non-paretic limb aligns with studies of other populations with unilateral impairments. For instance, in lower-limb amputees, the intact limb consistently shows higher loading than the prosthetic limb [40], with studies reporting twice the prevalence of knee pain and osteoarthritis in the intact knee compared to the general population [9]. Similarly, individuals with unilateral hip osteoarthritis demonstrate increased loading in their contralateral knee [7], suggesting this may be a common compensatory mechanism in unilateral lower extremity pathologies. On the contrary, Marrocco et al. [23] argued that stroke patients load their knees evenly, since they found no significant differences in knee abductor moment between paretic and non-paretic limbs in chronic stroke survivors. However, some participants showed increased loading on the paretic side, others on the non-paretic side, and some exhibited bilateral increases, when compared to healthy individuals. This discrepancy between joint moments and internal joint forces showed that individual stroke survivors may adopt different loading strategies between legs, something that was evident in our study. Our findings provide direct evidence of joint loading patterns that could not be fully captured by external moment analysis alone, highlighting the value of musculoskeletal modeling in understanding post-stroke gait adaptations.

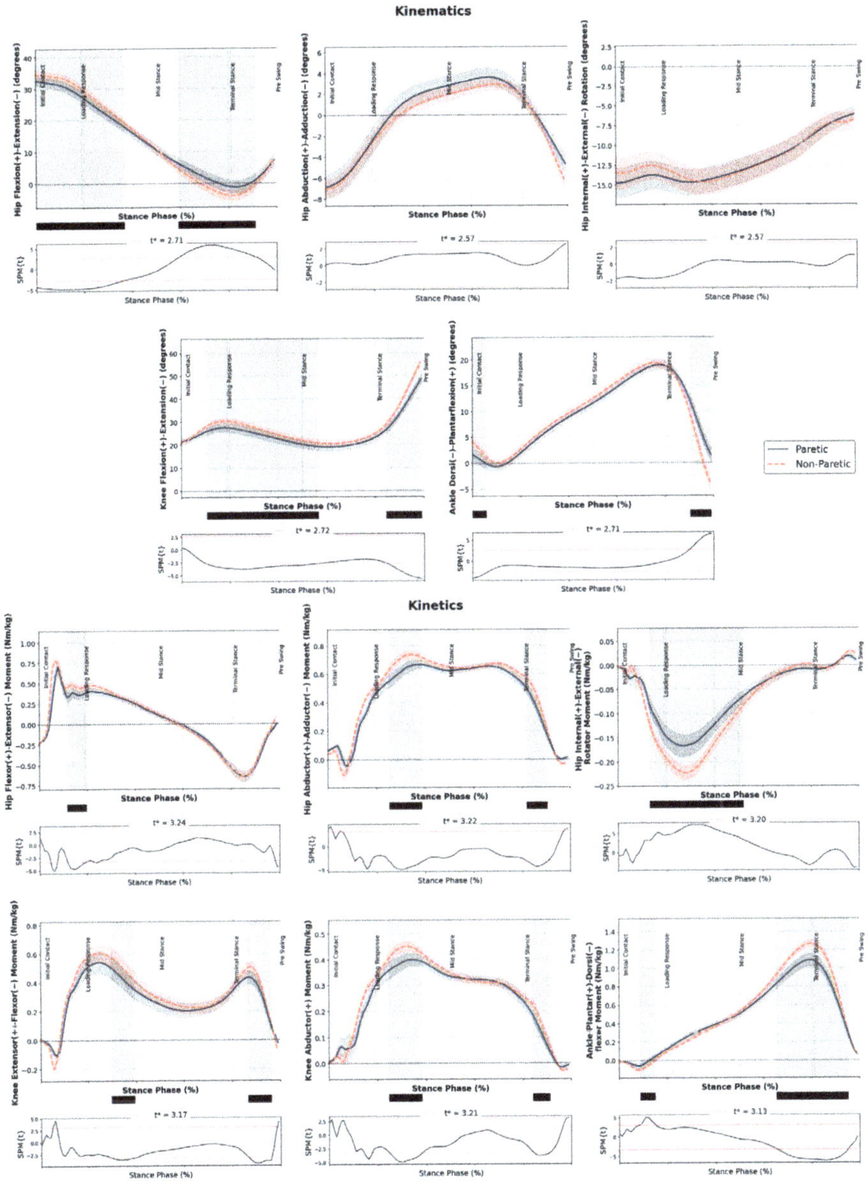

Figure 3. Lower limb kinematics and kinetics during stance phase for paretic (solid blue line) and non-paretic (dashed red line) limbs. The top panels show joint angles for hip (flexion–extension, abduction–adduction, and internal–external rotation), knee (flexion–extension), and ankle (dorsi–plantarflexion). The bottom panels show corresponding joint moments. Angles are in degrees and moments are normalized to body mass (Nm/kg). Shaded areas (red and black) represent 95% confidence intervals. Gray-shaded regions indicate periods of significant differences between limbs ($p < 0.05$) as determined by statistical parametric mapping. Black bars along the x-axis highlight the temporal regions of significant differences Horizontal dotted lines indicate zero levels and vertical dotted lines denote stance phase events: initial contact, loading response, mid-stance, terminal stance, and pre-swing.

Our regression analysis reveals important distinctions in how joint moments relate to knee contact forces between paretic and non-paretic limbs. Before discussing the results, we have to underline the differential effect of the knee moments on knee loading at frontal and fore–aft planes of motion. Knee abductor moment primarily loads the medial compartment through direct mechanical effect, by pulling the femur and tibia closer from the medial side, although passive elements, like the lateral collateral ligament, and muscles, like the tensor fasciae latae (through the iliotibial band), can also abduct the tibia. On the contrary, knee extensor moment indirectly affects overall knee loading, via the muscle force output used to counteract it. In the non-paretic limb, the first peak medial contact force showed a strong relationship with knee extensor moment ($R^2 = 0.573$), enhanced by the addition of knee abductor moment (final $R^2 = 0.671$). This finding differs from previous studies in knee osteoarthritis patients, where the knee abductor moment typically shows the strongest correlation with medial compartment loading [11]. The increased step width seen in stroke patients [41,42] to augment the base of support during walking could place the GRF vector closer to the knee joint center in the frontal plane, thus decreasing knee abductor moment and possibly its role to the medial knee joint force, as seen in OA patients walking during increased step width [11]. Interestingly, the paretic limb demonstrated an opposite pattern, with knee abductor moment being the primary predictor of first peak medial contact force ($R^2 = 0.559$). The latter more closely resembles the relationships reported in the osteoarthritis literature [10], suggesting that less flexed paretic knee allowed the corresponding moment to have a direct effect on medial loading. The weaker correlations observed for second peak forces in both limbs ($R^2 = 0.167$ paretic, $R^2 = 0.109$ non-paretic) align with previous research showing poor predictive value of joint moments for late stance loading [11], suggesting that muscle forces in multiple planes of motion relate to late stance knee loading.

The distinct moment–force relationships between limbs have important clinical implications. Traditional gait analysis focusing solely on inverse dynamics-based joint moments may not adequately capture the actual joint loading conditions in stroke survivors, particularly in the non-paretic limb where knee extensor moment plays a more prominent role than is typically assumed. Furthermore, the increased loading in the non-paretic limb, combined with altered movement patterns, may explain the higher risk of developing knee osteoarthritis in the non-paretic limb reported in longitudinal studies [28]. These findings suggest that rehabilitation strategies should consider not only restoring paretic limb function but also addressing the potentially harmful compensatory mechanisms in the non-paretic limb. Future interventions might benefit from incorporating targeted approaches to optimize load distribution between limbs while maintaining functional mobility.

Several limitations should be considered when interpreting our findings. First, the musculoskeletal model used in this study, while modified appropriately for knee contact force estimation, has not been directly validated against instrumented knee implant data in stroke patients. The model's accuracy in predicting knee contact forces depends on numerous assumptions regarding subject-specific knee geometry, muscle parameters, joint kinematics, and load distribution mechanisms, such as the absence of tendons or ligaments, which may differ in post-stroke gait compared to healthy or osteoarthritic individuals. Second, muscle activation patterns, which significantly influence joint contact forces, were estimated through static optimization, which minimizes muscle activations without accounting for the altered neuromuscular control and potential muscle co-contractions commonly observed in stroke survivors. Third, our model used generic rather than subject-specific parameters for muscle properties and joint alignments, which may impact force predictions, particularly in a population with potential muscle adaptations following stroke. Additionally, we did not explicitly quantify the potential impact of measurement noise from motion capture and force plate data on our joint loading estimates, though our use of appropriate filtering techniques (6 Hz low-pass filter) and established OpenSim processing pipelines should minimize noise-related errors in the final musculoskeletal modeling outcomes. Finally, the medial–lateral force distribution relies on simplified tibiofemoral geometry that cannot fully capture individual anatomical variations or potential joint deformities. Nevertheless, knee joint forces reported in this study align well with data from other modelling studies [43,44] and in vivo data from instrumented prostheses [45,46], indirectly validating our findings. These limitations highlight the need for future validation studies using instrumented implants or alternative measurement techniques in stroke populations to improve the accuracy of knee loading estimates.

In conclusion, this study provides strong evidence that musculoskeletal modeling offers superior insights into post-stroke joint mechanics compared to traditional biomechanical analysis approaches. The identified asymmetries in knee joint loading have important implications across several domains, as follows:

- **Clinical understanding:** The findings reveal biomechanical mechanisms underlying increased osteoarthritis risk in the non-paretic limb, moving beyond epidemiological observations to provide mechanistic insights into harmful joint loading patterns.
- **Methodological advancement:** The study establishes musculoskeletal modeling as a valuable clinical tool providing insights unavailable through traditional external moment analysis alone.
- **Clinical applications:** Results demonstrate that interventions should target both limbs simultaneously rather than focusing exclusively on paretic limb restoration, as compensatory loading patterns may lead to long-term consequences. These findings provide a foundation for developing interventions that address both functional recovery and long-term joint health preservation, potentially reducing secondary musculoskeletal complications in stroke survivors.

Author Contributions: Conceptualization, G.G., S.F. and N.A.; methodology, G.G. and M.M.; software, G.G.; validation, G.G., E.M. and E.G.; formal analysis, S.F.; resources, N.A. and K.V.; data curation, G.G., M.M., E.M. and S.F.; writing—original draft preparation, G.G. and E.G.; writing—review and editing, G.G., M.M., S.F., N.A., E.M. and K.V.; visualization, G.G.; supervision, N.A., K.V. and J.L.; project administration, N.A., E.G. and J.L.; funding acquisition, N.A. and K.V. All authors have read and agreed to the published version of the manuscript.

Funding: This work was funded by the project "Study of the interrelationships between neuroimaging, neurophysiological and biomechanical biomarkers in stroke rehabilitation" (MIS 5047286), which was implemented under the Action "Support for Regional Excellence" and funded by the Operational Program "Competitiveness, Entrepreneurship and Innovation" (NSRF 2014–2020) and co-financed by Greece and the European Union (European Regional Development Fund).

Institutional Review Board Statement: This study received ethical approval from the Research Ethics Committee of the Democritus University of Thrace (DUTH/EHDE/28061/165, 20 January 2023) and was conducted in accordance with international ethical rules.

Informed Consent Statement: Informed consent was obtained from all subjects involved in the study.

Data Availability Statement: The data presented in this study are available upon request from the corresponding author. The data are not publicly available due to privacy reasons.

Conflicts of Interest: The authors declare no conflicts of interest.

Appendix A

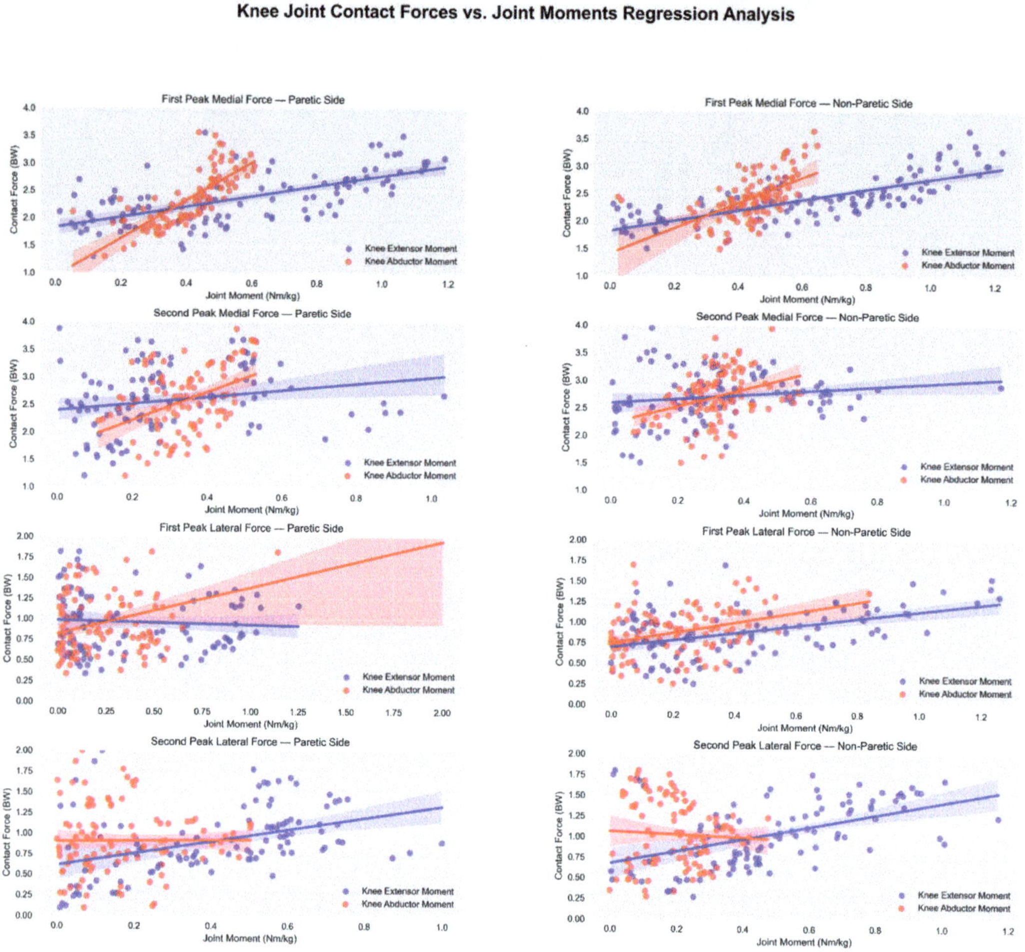

Figure A1. Regression analysis of knee joint contact forces and knee joint moments in post-stroke gait. Scatter plots showing the relationships between knee contact forces (y-axis, in body weight units) and joint moments (x-axis, in Nm/kg) for both paretic and non-paretic limbs. Blue points and lines represent knee extensor moments, while red points and lines represent knee abductor moments. Top panels show medial compartment forces, with first peak (early stance) and second peak (late stance) represented in rows 1 and 2, respectively. Bottom panels show lateral compartment forces, with first and second peaks in rows 3 and 4, respectively. Shaded areas represent confidence intervals of the regression estimates.

References

1. Wafa, H.A.; Wolfe, C.D.A.; Emmett, E.; Roth, G.A.; Johnson, C.O.; Wang, Y. Burden of Stroke in Europe: Thirty-Year Projections of Incidence, Prevalence, Deaths, and Disability-Adjusted Life Years. *Stroke* **2020**, *51*, 2418–2427. [CrossRef] [PubMed]
2. Johnson, C.O.; Nguyen, M.; Roth, G.A.; Nichols, E.; Alam, T.; Abate, D.; Abd-Allah, F.; Abdelalim, A.; Abraha, H.N.; Abu-Rmeileh, N.M.; et al. Global, Regional, and National Burden of Stroke, 1990–2016: A Systematic Analysis for the Global Burden of Disease Study 2016. *Lancet Neurol.* **2019**, *18*, 439–458. [CrossRef] [PubMed]
3. Belda-Lois, J.-M.; Horno, S.M.; Bermejo-Bosch, I.; Moreno, J.C.; Pons, J.L.; Farina, D.; Iosa, M.; Molinari, M.; Tamburella, F.; Ramos, A.; et al. Rehabilitation of Gait after Stroke: A Top down Approach. *J. Neuroeng. Rehabil.* **2011**, *8*, 66. [CrossRef]
4. Raghavan, P. Emerging Therapies for Spastic Movement Disorders. *Phys. Med. Rehabil. Clin. N. Am.* **2018**, *29*, 633–644. [CrossRef]

5. Luque-Moreno, C.; Ferragut-Garcías, A.; Rodríguez-Blanco, C.; Heredia-Rizo, A.M.; Oliva-Pascual-Vaca, J.; Kiper, P.; Oliva-Pascual-Vaca, Á. A Decade of Progress Using Virtual Reality for Poststroke Lower Extremity Rehabilitation: Systematic Review of the Intervention Methods. *BioMed Res. Int.* **2015**, *2015*, 342529. [CrossRef] [PubMed]
6. Cruz, T.H.; Lewek, M.D.; Dhaher, Y.Y. Biomechanical Impairments and Gait Adaptations Post-Stroke: Multi-Factorial Associations. *J. Biomech.* **2009**, *42*, 1673–1677. [CrossRef]
7. Shakoor, N.; Foucher, K.C.; Wimmer, M.A.; Mikolaitis-Preuss, R.A.; Fogg, L.F.; Block, J.A. Asymmetries and Relationships between Dynamic Loading, Muscle Strength, and Proprioceptive Acuity at the Knees in Symptomatic Unilateral Hip Osteoarthritis. *Arthritis Res. Ther.* **2014**, *16*, 455. [CrossRef]
8. Andriacchi, T.P.; Mündermann, A. The Role of Ambulatory Mechanics in the Initiation and Progression of Knee Osteoarthritis. *Curr. Opin. Rheumatol.* **2006**, *18*, 514–518. [CrossRef]
9. Robbins, C.B.; Vreeman, D.J.; Sothmann, M.S.; Wilson, S.L.; Oidridge, N.B. A Review of the Long-Term Health Outcomes Associated With War-Related Amputation. *Mil. Med.* **2009**, *174*, 588–592. [CrossRef]
10. Holder, J.; van Drongelen, S.; Uhlrich, S.D.; Herrmann, E.; Meurer, A.; Stief, F. Peak Knee Joint Moments Accurately Predict Medial and Lateral Knee Contact Forces in Patients with Valgus Malalignment. *Sci. Rep.* **2023**, *13*, 2870. [CrossRef]
11. Richards, R.E.; Andersen, M.S.; Harlaar, J.; van den Noort, J.C. Relationship between Knee Joint Contact Forces and External Knee Joint Moments in Patients with Medial Knee Osteoarthritis: Effects of Gait Modifications. *Osteoarthr. Cartil.* **2018**, *26*, 1203–1214. [CrossRef] [PubMed]
12. Walter, J.P.; D'Lima, D.D.; Colwell, C.W.; Fregly, B.J. Decreased Knee Adduction Moment Does Not Guarantee Decreased Medial Contact Force during Gait. *J. Orthop. Res.* **2010**, *28*, 1348–1354. [CrossRef] [PubMed]
13. Baert, I.A.C.; Jonkers, I.; Staes, F.; Luyten, F.P.; Truijen, S.; Verschueren, S.M.P. Gait Characteristics and Lower Limb Muscle Strength in Women with Early and Established Knee Osteoarthritis. *Clin. Biomech.* **2013**, *28*, 40–47. [CrossRef]
14. Chen, G.; Patten, C.; Kothari, D.H.; Zajac, F.E. Gait Differences between Individuals with Post-Stroke Hemiparesis and Non-Disabled Controls at Matched Speeds. *Gait Posture* **2005**, *22*, 51–56. [CrossRef]
15. Buurke, J.H.; Nene, A.V.; Kwakkel, G.; Erren-Wolters, V.; Ijzerman, M.J.; Hermens, H.J. Recovery of Gait after Stroke: What Changes? *Neurorehabil. Neural Repair* **2008**, *22*, 676–683. [CrossRef]
16. Allen, J.L.; Kautz, S.A.; Neptune, R.R. Forward Propulsion Asymmetry Is Indicative of Changes in Plantarflexor Coordination during Walking in Individuals with Post-Stroke Hemiparesis. *Clin. Biomech.* **2014**, *29*, 780–786. [CrossRef]
17. Robbins, S.M.K.; Maly, M.R. The Effect of Gait Speed on the Knee Adduction Moment Depends on Waveform Summary Measures. *Gait Posture* **2009**, *30*, 543–546. [CrossRef]
18. Yang, C.P.; Lee, C.L.; Chen, T.W.; Lee, S.; Weng, M.C.; Huang, M.H. Ultrasonographic Findings in Hemiplegic Knees of Stroke Patients. *Kaohsiung J. Med. Sci.* **2005**, *21*, 70–77. [CrossRef] [PubMed]
19. Aderibigbe, A.S.; Famurewa, O.C.; Komolafe, M.A.; Omisore, A.D.; Adetiloye, V.A. Sonographic Soft Tissue Arthritic Changes Associated with Post-Stroke Hemiplegic Knee Pain: Utility of Musculoskeletal Ultrasound in a Resource-Limited Setting. *Polish J. Radiol.* **2020**, *85*, e45. [CrossRef]
20. Tunç, H.; Öken, Ö.; Kara, M.; Tiftik, T.; Doğu, B.; Ünlü, Z.; Özçakar, L. Ultrasonographic Measurement of the Femoral Cartilage Thickness in Hemiparetic Patients after Stroke. *Int. J. Rehabil. Res.* **2012**, *35*, 203–207. [CrossRef]
21. Jeong, K.Y.; Lee, H.J. Prevalence of Knee Osteoarthritis and Health-Related Quality of Life in Stroke Patients over 60 Years Old: A Cross-Sectional Study Using Korean National Health and Nutrition Examination Survey V. *Ann. Geriatr. Med. Res.* **2021**, *25*, 178. [CrossRef]
22. Fournier, J.; Finestone, H.; Lauzon, J.; Campbell, T.M. Prevalence, Impact, and Treatment of Co-Occurring Osteoarthritis in Patients With Stroke Undergoing Rehabilitation: A Review. *Stroke* **2021**, *52*, E618–E621. [CrossRef]
23. Marrocco, S.; Crosby, L.D.; Jones, I.C.; Moyer, R.F.; Birmingham, T.B.; Patterson, K.K. Knee Loading Patterns of the Non-Paretic and Paretic Legs during Post-Stroke Gait. *Gait Posture* **2016**, *49*, 297–302. [CrossRef] [PubMed]
24. Shen, K.H.; Prajapati, S.K.; Borrelli, J.; Gray, V.L.; Westlake, K.P.; Rogers, M.W.; Hsiao, H.Y. Neuromechanical Control of Impact Absorption during Induced Lower Limb Loading in Individuals Post-Stroke. *Sci. Rep.* **2022**, *12*, 19104. [CrossRef]
25. Holder, J.; Trinler, U.; Meurer, A.; Stief, F. A Systematic Review of the Associations Between Inverse Dynamics and Musculoskeletal Modeling to Investigate Joint Loading in a Clinical Environment. *Front. Bioeng. Biotechnol.* **2020**, *8*, 1382. [CrossRef] [PubMed]
26. Mohout, I.; Elahi, S.A.; Esrafilian, A.; Killen, B.A.; Korhonen, R.K.; Verschueren, S.; Jonkers, I. Signatures of Disease Progression in Knee Osteoarthritis: Insights from an Integrated Multi-Scale Modeling Approach, a Proof of Concept. *Front. Bioeng. Biotechnol.* **2023**, *11*, 1214693. [CrossRef] [PubMed]
27. Asghari, M.; Behzadipour, S.; Taghizadeh, G. A Planar Neuro-Musculoskeletal Arm Model in Post-Stroke Patients. *Biol. Cybern.* **2018**, *112*, 483–494. [CrossRef]
28. Andersen, M.S. Introduction to Musculoskeletal Modelling. In *Computational Modelling of Biomechanics and Biotribology in the Musculoskeletal System: Biomaterils and Tissues*; Woodhead Publishing: Cambridge, UK, 2021; pp. 41–80.

29. Giarmatzis, G.; Fotiadou, S.; Giannakou, E.; Kokkotis, C.; Fanaradelli, T.; Kordosi, S.; Vadikolias, K.; Aggelousis, N. Understanding Post-Stroke Movement by Means of Motion Capture and Musculoskeletal Modeling: A Scoping Review of Methods and Practices. *BioMed* **2022**, *2*, 409–421. [CrossRef]
30. Curreli, C.; Di Puccio, F.; Davico, G.; Modenese, L.; Viceconti, M. Using Musculoskeletal Models to Estimate in Vivo Total Knee Replacement Kinematics and Loads: Effect of Differences Between Models. *Front. Bioeng. Biotechnol.* **2021**, *9*, 703508. [CrossRef]
31. Tomasi, M.; Artoni, A.; Mattei, L.; Di Puccio, F. On the Estimation of Hip Joint Loads through Musculoskeletal Modeling. *Biomech. Model. Mechanobiol.* **2022**, *22*, 379–400. [CrossRef]
32. D'Lima, D.D.; Fregly, B.J.; Patil, S.; Steklov, N.; Colwell, C.W. Knee Joint Forces: Prediction, Measurement, and Significance. *Proc. Inst. Mech. Eng. Part H J. Eng. Med.* **2012**, *226*, 95–102. [CrossRef]
33. Giarmatzis, G.; Fotiadou, S.; Giannakou, E.; Karakasis, E.; Vadikolias, K.; Aggelousis, N. Evaluating the Repeatability of Musculoskeletal Modelling Force Outcomes in Gait among Chronic Stroke Survivors: Implications for Contemporary Clinical Practice. *Biomechanics* **2024**, *4*, 333–345. [CrossRef]
34. Lerner, Z.F.; DeMers, M.S.; Delp, S.L.; Browning, R.C. How Tibiofemoral Alignment and Contact Locations Affect Predictions of Medial and Lateral Tibiofemoral Contact Forces. *J. Biomech.* **2015**, *48*, 644–650. [CrossRef]
35. Delp, S.L.; Anderson, F.C.; Arnold, A.S.; Loan, P.; Habib, A.; John, C.T.; Guendelman, E.; Thelen, D.G. OpenSim: Open-Source Software to Create and Analyze Dynamic Simulations of Movement. *IEEE Trans. Biomed. Eng.* **2007**, *54*, 1940–1950. [CrossRef] [PubMed]
36. Pataky, T.C. One-Dimensional Statistical Parametric Mapping in Python. *Comput. Methods Biomech. Biomed. Engin.* **2012**, *15*, 295–301. [CrossRef] [PubMed]
37. Pataky, T.C. Www.Spm1d.Org. Available online: https://spm1d.org/ (accessed on 26 May 2025).
38. Seabold, S.; Perktold, J. Statsmodels: Econometric and Statistical Modeling with Python. In Proceedings of the 9th Python in Science Conference, Austin, TX, USA, 28 June–3 July 2010; pp. 92–96.
39. Liu, T.; Xie, H.; Yan, S.; Zeng, J.; Zhang, K.; Liu, T.; Xie, H.; Yan, S.; Zeng, J.; Zhang, K. The Effect of Thigh Muscle Forces on Knee Contact Force in Female Patients with Severe Knee Osteoarthritis. *Bioengineering* **2024**, *11*, 1299. [CrossRef]
40. Miller, R.H.; Krupenevich, R.L.; Pruziner, A.L.; Wolf, E.J.; Schnall, B.L. Medial Knee Joint Contact Force in the Intact Limb during Walking in Recently Ambulatory Service Members with Unilateral Limb Loss: A Cross-Sectional Study. *PeerJ* **2017**, *2017*, e2960. [CrossRef] [PubMed]
41. Hak, L.; Houdijk, H.; Van Der Wurff, P.; Prins, M.R.; Mert, A.; Beek, P.J.; Van Dieën, J.H. Stepping Strategies Used by Post-Stroke Individuals to Maintain Margins of Stability during Walking. *Clin. Biomech.* **2013**, *28*, 1041–1048. [CrossRef]
42. Hak, L.; Blokland, I.; Houdijk, H. How Do People after Stroke Adapt Step Parameters and Margins of Stability at Different Walking Speeds? *Gait Posture* **2020**, *81*, 136–137. [CrossRef]
43. Kumar, D.; Manal, K.T.; Rudolph, K.S. Knee Joint Loading during Gait in Healthy Controls and Individuals with Knee Osteoarthritis. *Osteoarthr. Cartil.* **2013**, *21*, 298–305. [CrossRef]
44. Giarmatzis, G.; Zacharaki, E.I.; Moustakas, K. Real-Time Prediction of Joint Forces by Motion Capture and Machine Learning. *Sensors* **2020**, *20*, 6933. [CrossRef] [PubMed]
45. Kutzner, I.; Heinlein, B.; Graichen, F.; Bender, A.; Rohlmann, A.; Halder, A.; Beier, A.; Bergmann, G. Loading of the Knee Joint during Activities of Daily Living Measured In Vivo in Five Subjects. *J. Biomech.* **2010**, *43*, 2164–2173. [CrossRef] [PubMed]
46. Lima, D.D.D.; Patil, S.; Steklov, N.; Slamin, J.E.; Jr, C.W.C. Tibial Forces Measured In Vivo After Total Knee Arthroplasty. *J. Arthroplast.* **2006**, *21*, 255–262. [CrossRef] [PubMed]

biomechanics

Article

Effects of Simulated Hyper-Gravity on Lower Limb Kinematics and Electromyography During Walking

Christopher A. Malaya [1,2,*], Pranav J. Parikh [2], Dean L. Smith [3] and Charles S. Layne [2]

[1] Grail Laboratory, Parker University, Dallas, TX 75229, USA
[2] Center for Neuromotor and Biomechanics Research, Department of Health and Human Performance, University of Houston, Houston, TX 77204, USA; pjparikh2@uh.edu (P.J.P.); clayne2@central.uh.edu (C.S.L.)
[3] Department of Kinesiology, Nutrition, and Health, Miami University, Oxford, OH 45056, USA; smithdl2@miamioh.edu
* Correspondence: cmalaya@parker.edu

Abstract: Background: Gravity profoundly influences human locomotion. Studies examining the effects of hyper-gravity on gait have largely relied on added external mass, potentially confounding results with changes in inertia and center of mass. This study attempted to isolate the effects of increased gravitational load on kinematics and electromyography during walking at several different levels of load. **Methods:** Fifteen healthy adults were exposed to simulated gravitational loads ranging from 100% to 130% of body weight using a novel harness and spring-based system that increased weight without the addition of external mass and without altering limb inertia. Participants walked on a treadmill at a self-selected speed through incremental loading and unloading. Lower limb kinematics and electromyography data were recorded. Traditional measures of gait, as well as more dynamical measures, including angle–angle analysis and phase portraits, were examined. **Results:** Data demonstrated that a 130% load is sufficient to induce kinematic changes at the hip and knee; however, these changes become significant only during the transition from 130% to lower load levels. Ankle kinematics and electromyography appeared to be unaffected. **Conclusions:** These findings suggest that the presence of external mass and alterations in limb inertias should be considered seriously as independent variables in future loading studies, and that weight and mass may need to be considered as separate effectors during locomotion. This study also found that the act of loading and unloading elicit distinct responses in the joints of the lower extremities, as well as that it may induce an adaptative after-effect.

Keywords: gait; gravity; loading; hysteresis; kinematics; EMG; treadmill

1. Introduction

Gravity is a ubiquitous natural phenomenon that pervades every aspect of human experience on Earth. While the acceleration caused by gravity (g) varies from 9.763 to 9.833 m per second squared (m/s^2) depending on your terrestrial location, "standard g" is often modeled—and assumed—at 9.81 m/s^2 [1]. Though humans experience this acceleration as their own weight—the force borne of their own mass accelerating—the influence of our gravitational environment extends far beyond perception and weightiness. Gravity guides the formation of human bone structure and density [2], influences the discharge rates and amplitudes of cortical and spinal neurons [3], and can alter cellular morphology and metabolism [4]. However, there is arguably no system, structure or behavior affected so demonstrably by gravity as that of human locomotion.

Locomotion studies utilizing environments with increased gravitational effects (hyper-gravity) are extremely limited. Previous work investigating the effects of increased weight on human locomotion have found changes in trunk angles, hip, and ankle range of motion and cadence [5], as well as equilibrium points, stability, efficiency, gait speed, and walk-to-run transitions [6–8]. However, the studies contained by these reviews utilized the addition of mass to participant's bodies in order to increase weight. While the goal of these studies was to examine the effects of externally carried loads, it highlights an important confound as these—and similar studies—relate to the effects of hyper-gravity. The addition of external mass changes an individual's center of mass location [9], induces a stabilizing forward lean [10,11], and adds impactful, unevenly distributed inertial differences to a body [12,13]. These factors, while related to mass and weight, may be unrelated or tangential to purely gravitational constraints. Therefore, the extent to which these changes are driven by increased weight versus increased inertia on the limbs and trunk remains unknown.

Similarly, it is unclear if the kinematic and electromyographic changes seen during loading are due entirely to the relative load on an individual, or if the previous level of load may work as a factor in these responses. In humans, hysteresis—the dependence of a system on its previous states—has been a dynamic factor exhibited at the cellular level [14–17], scaling all the way up to cortical networks [18,19], proprioception [20,21] and gait [22–26].

It is possible that the same level of load could elicit different kinematic and EMG changes, depending on whether an individual increased or decreased their relative weight to reach it. However, very few studies have sought to examine hysteresis in human gait through manipulation of the gravitational load. Without a clear understanding as to the specific and exact role of loading and unloading, as well as specific increases in load and mass in these adaptations, it is difficult, if not impossible, to optimize load carriage for real world conditions and outcomes. In this study, we sought to investigate the effects of increased gravitational loading on kinematic and electromyographic variables without the addition of external mass. Specifically, we were interested in examining the kinematics and neuromuscular activity of individuals in simulated gravitational environments ranging from 100%, up to 130% of gravitational load. In this study, these questions were approached through the use of zero-dimensional (traditional kinematic and electromyographic measures) and one-dimensional (angle-angle diagrams and phase portraits) analyses. Previous use of these methods in this lab has found that they provide complementary information not otherwise apparent given the use of a single set of measures [27,28].

2. Methods

2.1. Participants

This study examined 15 healthy adults (25.3 ± 4.7 years; 67.1 ± 4.0 inches; 172.3 ± 42.0 lbs.; 53% female). Participants were not knowingly pregnant and did not have a history of, or any current systemic, degenerative or neuromusculoskeletal injuries or disease that could affect their ability to walk with differential loading for 15 min. This study was conducted in accordance with the Declaration of Helsinki and approved by the Institutional Review Board at the University of Houston (IRB#:00002971). Informed consent was obtained from all participants prior to enrollment in the study.

2.2. Kinematic Sensors

Participants were fitted with seven inertial measurement units (IMUs; XSens Awinda—Movella Inc., Henderson, NV, USA) arranged in a lower-body configuration. These sensors were placed bilaterally over the insteps of both feet, as well as anteriorly over the tibia at mid-shank and laterally over the mid-thigh. The final sensor was placed over the sacrum, centered at the S2 tubercle. All XSens sensors were secured by proprietary neoprene straps with non-slip, rubber backings.

2.3. Electromyographic Sensors

Four dry surface electromyographic (EMG) sensors (Model SX230—Biometrics Ltd., Newport, RI, USA) were adhered—using hypo-allergenic, double-sided tape—over the right rectus femoris, biceps femoris, medial gastrocnemius and tibialis anterior. These sensors were placed over the belly of each respective muscle—conduction surfaces in line with the muscle fibers—after any body hair in that location was shaved, and the area was cleaned and scrubbed with an alcohol wipe. Sensor placements were performed in accordance with recommendations by the Surface Electromyography for the Non-Invasive Assessment of Muscles (SENIAM) group. The control system (DataLOG MWX8—Biometrics Ltd., Newport, RI, USA) was mounted on each participant's low back using a stretchable, Velcro band around their waist.

2.4. Loading System and Walking Protocol

Participants were asked to wear a climbing harness with front and back D-ring attachments over their clothing. This harness allowed the participants to be attached to the loading system at two points of equal height, thereby creating an equivalent angle in the front and back ropes that would tether them to the system. This had the intended effect of canceling out any anterior or posterior forces from the system, leaving only a vertical component of load. Similarly, the loading system was connected to the harness by two springs, which allowed individuals in a small degree of normal displacement that would not be available if they were only connected to a taut rope. The entire harness, after being connected to the rope system, added 4 lbs. of weight distributed over the shoulders of the participant (see Figure 1). The entire loading system was built around a treadmill with embedded force plates (Bertec Corporation, Columbus, OH, USA); as such, after donning the harness and being connected to the loading system, an individual's weight could be calculated and loading parameters for 100%, 110%, 120%, and 130% of normal body weight were established.

Figure 1. Loading system. In this figure, the participant (light blue humanoid) walks over an instrumented treadmill (green). They are connected to the rope system (black) by two springs (silver) that attach to a harness (not pictured). The ropes distribute vertical tension by way of 8 pulleys (orange) arranged around a metal frame (red). The tension in the rope system can be modulated by way of a crank pulley (dark blue) and vertical load is calculated by kinetic sensors embedded in the treadmill.

Participants were asked to walk at 100% of their normal load for 5 min at a self-selected, comfortable speed (mean speed: 0.78 ± 0.11 m/s). This gave each participant time to become familiar with the loading system, as well as for their gait to stabilize. Following the acclimation period, participants were loaded in 10% increments up to 130% of their body weight, spending 1 full minute at each level. After completing the full minute at 130% of body weight, the protocol was reversed, with participants walking for one minute at 120%, 110%, and 100% of normal load. For all levels of loading and unloading, the treadmill speed remained at the same speed each participant had selected earlier. There was no rest given in between each level besides the time it took to adjust the system to the desired load (≈10 s). Kinematic and EMG data were recorded for the final minute of the acclimatization period, as well as the full minute of walking at all levels of loading and unloading.

2.5. Data Processing

Kinematic data were streamed wirelessly from the XSens IMUs at 60 Hz to a computer running a data collection software suite (MVN Awinda ver. 2022.1). This software collected and internally calculated joint angles for the hip, knee, and ankle, bilaterally. Joint angle waveforms were separated into strides and normalized to 100 points using the peak knee as a reference. Mean, maximum, and minimum angles, as well as range of motion (ROM) were extracted for all joints. Data were exported, separated into strides using peak knee as the reference point, and statistically analyzed in MATLAB (R2019b: 9.7.0.1296695) using custom scripting.

Four channels of EMG data were simultaneously recorded by the waist-mounted control unit, as well as streamed to a computer running a data collection software suite (DataLOG ver: 10.27—Biometrics Ltd., Newport, RI, USA). The 12 g DataLOG control unit was set to sample at 1000 Hz, and provided 1000× amplification gain as well as an automatic anti-aliasing filter prior to streaming. Data collected were exported into MATLAB for processing. Each channel was individually bandpass filtered (20 to 450 Hz) using a 2nd order Butterworth filter. Waveforms were then full wave rectified and enveloped using a low pass filter with an additional 2nd order Butterworth filter utilizing a cutoff frequency of 40 Hz [29]. EMG data were separated into strides and normalized to 100 points using the kinematic peak knee timestamps as a reference. After processing, peak values, root-mean-square (RMS) and integrated areas were calculated for all muscles. We calculated RMS as the square root of the mean of all values squared for each trial. This provides a metric representing the amplitude of the EMG signal [30]. We also calculated integrated areas for each trial to appraise the total electrical signal or drive from the central nervous system to the motorneuron [31–34].

This study made use of both zero- and one-dimensional analyses, representing traditional kinematic and electromyographic measures as well as phase portraits and angle-angle diagrams. These were created to examine the state spaces of and coordination between the joints of the lower extremity, respectively. Areas were calculated from mean phase portraits using a custom MATLAB script in order to quantify and compare the range of available behaviors.

2.6. Statistical Analysis

Kinematic and electromyographic variables were tested for normality and sphericity using the Kolmogorov–Smirnov and Shapiro–Wilk tests, as well as Mauchly's test, respectively. Mean, maximum, minimum angles, and range of motion for each joint, as well as peak value, RMS, and integrated areas were compared across all levels of loading using repeated measure ANOVAs. Post hoc testing was performed with paired *t*-tests, as appropriate.

3. Results

Kolmogorov–Smirnov and Shapiro–Wilk tests revealed all data were normally distributed and Mauchly's test showed sphericity was preserved.

3.1. Kinematics

Hip, knee, and ankle average joint angle waveforms by loading level are presented in Figure 2. Results from repeated measures ANOVA showed loading level had a statistically significant effect on hip mean ($F(6,84) = 2.447$, $p = 0.0314$) and max ($F(6,84) = 3.073$, $p = 0.0091$) values, as well as knee mean ($F(6,84) = 3.172$, $p = 0.0074$), and min ($F(6,84) = 4.647$, $p = 0.0004$) values. Pairwise comparisons for the hip and knee are depicted in Table 1. No differences in the ankle variables or any ROM values were found to be significant.

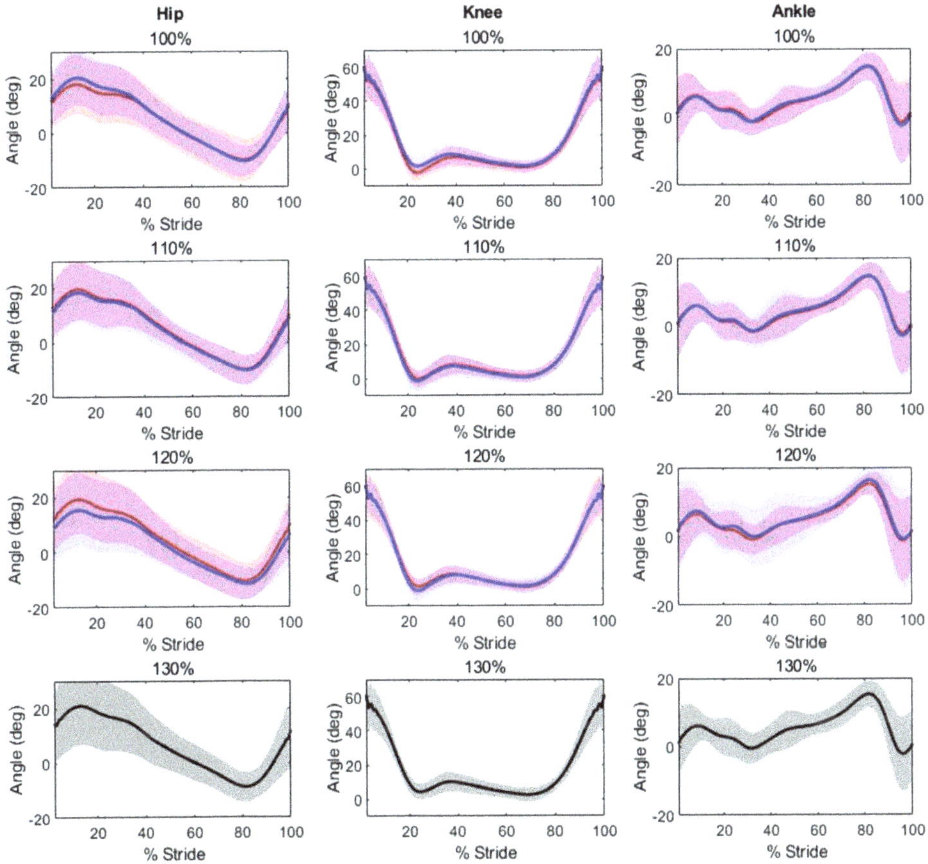

Figure 2. Hip, knee, and ankle joint angles by loading level. Each plot contains the kinematic waveforms for its respective loading (in red) and unloading (in blue) condition, along with a 2-standard deviation shaded area around each waveform. All 130% load conditions are in black to avoid any confusion, as only a single waveform is present.

Table 1. Hip and knee joint angles—pairwise comparisons.

		Condition	$\mu^{\circ} \pm std$	p Value
Hip	Mean	100	4.6 ± 7.6	0.1241
		110	5.2 ± 5.9	0.4919
		120	4.8 ± 8.6	0.0647
		130	6.4 ± 8.5	
		120 U	2.6 ± 8.3	0.0138 *
		110 U	4.5 ± 7.0	0.0642
		100 U	5.3 ± 6.5	0.3423
	Max	100	19.3 ± 10.9	0.0463 *
		110	20.5 ± 10.2	0.4931
		120	20.3 ± 12.1	0.1595
		130	22.0 ± 13.5	
		120 U	16.7 ± 12.9	0.0311 *
		110 U	19.2 ± 11.0	0.0488 *
		100 U	21.2 ± 9.8	0.6390
Knee	Mean	100	15.1 ± 5.1	0.0001 *
		110	16.2 ± 4.9	0.0068 *
		120	16.4 ± 4.6	0.0113 *
		130	17.8 ± 4.9	
		120 U	16.1 ± 3.8	0.1316
		110 U	15.4 ± 5.1	0.0082 *
		100 U	16.7 ± 4.7	0.1020
	Min	100	−3.1 ± 5.3	0.0012 *
		110	−1.7 ± 4.2	0.0055 *
		120	−1.1 ± 3.7	0.0027 *
		130	1.0 ± 3.9	
		120 U	−2.2 ± 6.3	0.0554
		110 U	−2.6 ± 5.6	0.0081 *
		100 U	−0.6 ± 4.8	0.0782

* denotes significance ($p < 0.05$). All pair-wise comparisons depicted represent the specified measure at that level of load versus 130% load. "U" denotes the specified level of load as it was unloaded to.

3.2. Electromyography

There were no significant differences in levels of load for peak muscle activity, root-mean-square, or integrated areas for any muscle. Mean and standard deviation EMG values by muscle, variable, and condition can be found in the Supplementary Materials.

4. Discussion

This study examined the effects of simulated gravitational loading (in this case, increased weight without the addition of extra, external mass) between 100% and 130% of body weight on kinematics and electromyographic variables during walking. We were interested in investigating if 130% of body weight was a sufficient load to induce kinematic and EMG changes, as well as examining the individual effects of increased weight without the addition of external mass. Our data revealed that 130% load is sufficient to elicit kinematic changes; however, these changes only appear significant when unloading from 130% to lesser loads. This suggests that walking at 130% and then unloading leads to gait alterations, while simply loading up to 130% does not. It is thus potentially the act of loading or unloading that can elicit changes at these levels, in addition to the actual borne weight.

Preceding studies have demonstrated that human proprioception diminishes in hypogravitational environments [35–37]. Indeed, anticipatory postural adjustments disappear below normal gravitational conditions [38] and kinesthetic responses to vibration diminish [39], with these changes being displayed not only kinematically, but also in the human cortical waveforms [40,41]. These studies indicate that alterations in human proprioception due to hypogravity are far reaching, and prevalent. This study found that increased loading at 130% of body weight was sufficient to elicit kinematic changes, but these changes were only clear as participants were unloaded to 120% of their body weight. To be clear, all levels of load from 100% increasing to 130% were not statistically distinct, yet 130% was significantly different from 120% in the knee and hip as participants were unloaded. This suggests that the level of load may not be the only operative factor in our findings; rather, the acts of loading or unloading may elicit distinct kinematic responses. Studies examining the drivers of hysteresis found that hysteretic effects were highest in the situations in which sensory information was the weakest [42] and that perceptual judgements are affected by the lack of or availability of information about an impending action [22].

In the case of this study, movement from a higher level of load (130%) to a lower level of load (120%) would reduce the relative amount of available sensory information. This, in turn, would invoke hysteretic changes in which participants based their expectation of movement in the new environment less on actual environmental cues, and instead more on internal models and expectation.

This concept is supported by Kostyukov and Cherkassky [43], which found muscle spindle discharge rates were higher after stimulation rate increases, and lower after decreases. It is also possible that some of these effects are modified by plantar pressure. Work by Kozlovskaya et al. [44] found that the removal of plantar support led to reflexive decreases in muscle activity and the eventual atony of extensor musculature with concurrent reductions in proprioception [45,46]. Exposure over longer time frames has led to decreased muscle strength-speed properties and motor control alterations [47,48]. Further, some of these alterations were entirely mitigated with plantar pressure stimulation [49]. As load increases, the relative increase in environmental-based proprioceptive information will drive gait behaviors more strongly; on the contrary, as the relative availability of proprioceptive information decreases, the reduction of sensory information will facilitate the use of information from previous levels of loading. This suggests that the effects of unloading, and loading are kinematically distinct, and should likely not be considered equivalent factors, even if used to reach the same level of load.

Phase portraits graphically represent all of the potential states of a dynamic system [50]. In this case, phase portraits depict all of the potential positions (i.e., angles) of a joint, as well as their velocities at that moment. A direct comparison of the portraits for 100% and 130% load (see Figure 3) shows that the movement structure of the joints is mostly preserved, with some stretching and translation as load increases. In combination with the angle-angle diagrams (see Figure 4), we can also see that the coordination between the joints is relatively similar, but also expanded and translated. This suggests that walking-type gait is relatively robust from 100% to 130% of body weight. Interestingly, this mirrors previous investigations of unloading down from full body weight and strengthens the idea that gait is a behavior centered around and suited to our particular gravitational environment. While this study did not present enough load to examine if an entirely new locomotive behavior would emerge at very high percentages of body weight (analogous and opposite to the sub-volitional shift into bounding-type gait found on the moon, for example), the durability of walking-type locomotion appears to be strong up to a 130% load.

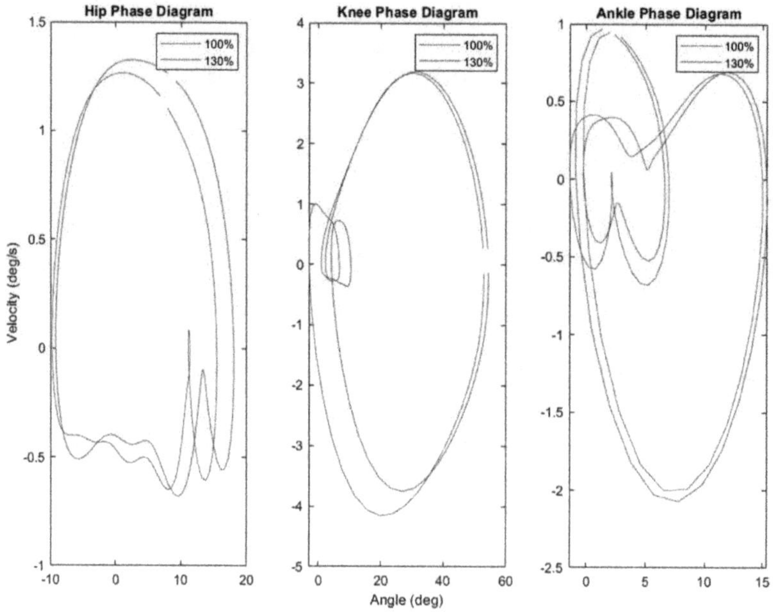

Figure 3. Phase portrait comparisons of 100% and 130% load. This graph displays phase portraits for the hip, knee, and ankle. Though there is some expansion of the range of available behaviors for the hip and knee—suggesting they are most sensitive to loading—the ankle appears to be mostly unaffected by the increase in load.

We also calculated the areas encircled by our phase portraits for every level of load by joint (see Figure 5). These areas are the two-dimensional spaces created by the outermost set of lines on each graph. These values provide a quantification of the state space of each joint at each level of load; in that way, they can be considered a means to numerically compare the contraction or expansion of the state space between different conditions. In this study, examination of the phase portrait areas of the knee reveals steady contraction of the state space as we increased to 130% of load, before a more than 10% expansion at 120% unloading. Interestingly, this expansion then contracts as we continue to unload, eventually settling at a smaller area than even the original (100%) load condition. The hip areas, on the contrary, consistently expand as we increase to a 120% load—drop slightly at 130%—before contracting significantly as we unload back to a 120% load. Analogously to the knee, the area of the hip phase portrait then continues increasing as load decreases, eventually ending at a larger area than even the original (100%) load condition. This suggests a crucially interesting relationship between the hip and the knee: as the range of available configurations of the hip expands, the knee, inversely, contracts. Similarly, as the hip contracts, the knee expands, and vice versa.

A consideration of the total areas across all three joints (see Table 2) finds a similar trend to the above. As loading increases to 130%, the overall summed areas of the three joints contract slightly before an almost 10% increase with unloading from 130% to 120%. This area remains relatively stable with unloading to 110% and then drops 4% with a return to 100% load. This, and the previous trending (in the hip and knee) highlight two primary ideas. First, this supports the previous assertion that loading and unloading do not appear to be equivalent phenomena. Second, while it is possible that there is an inflection point at 110% with unloading, it is also possible—given the similarity of 120% and 110% when unloading—that this is extinction of a loading induced after-effect. This has major implications for populations in which load-carriage is common in that effective increases or decreases in weight can alter kinematics and movement structure, possibly even for time beyond the actual adding or subtracting of weight. Correspondingly, whether the individual was loaded or unloaded to a certain weight appears to induce specific changes that are not equal across similar loading levels.

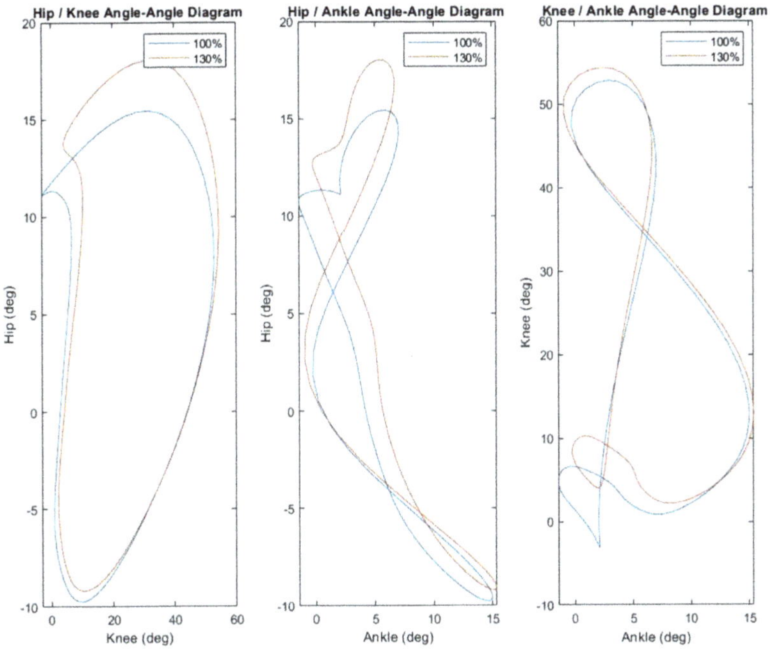

Figure 4. Angle-angle diagram comparisons of 100% and 130% load. This graph shows that the coordination strategies between the joints of the lower extremities are mostly preserved as load was increased. It is important to note, however, that there was distinct stretching and skewing of the shapes for all three graphs. This suggests that, although the general coordinative structure of movement between these joints was similar, they were not unaffected by load. Indeed, even 130% of body weight was enough to shift some aspects of the coordinative structure of the hip, knee, and ankle.

Interestingly, the kinematic changes seen in this study were not reflected in EMG data, in which no changes were found across any levels of load. This suggests that muscle activity and kinematic variables can decouple and respond to changes in load differentially. While previous investigations have found that kinematics can be accurately predicted from EMG data alone [51], others have found that kinematic and EMG variables correlate differentially depending on the activity being performed [52]. The results of this study suggest that loading and unloading are activities in which these variables do not track well with each other. It is important to note that this is also potentially due the absence of external mass. In this study, changes in limb inertia and center of gravity were bypassed through the use of our novel loading system. In this way, and in relation to load added over body weight, kinematic variables appear more sensitive than EMG to loading and unloading, and perhaps—given the results of other studies with positive EMG findings—EMG is more sensitive to changes stemming from changes in external mass/inertia. Alternatively, as we only recorded EMG data from four muscles, it is possible that other muscles exhibited changes in response to the loading protocol but were not captured.

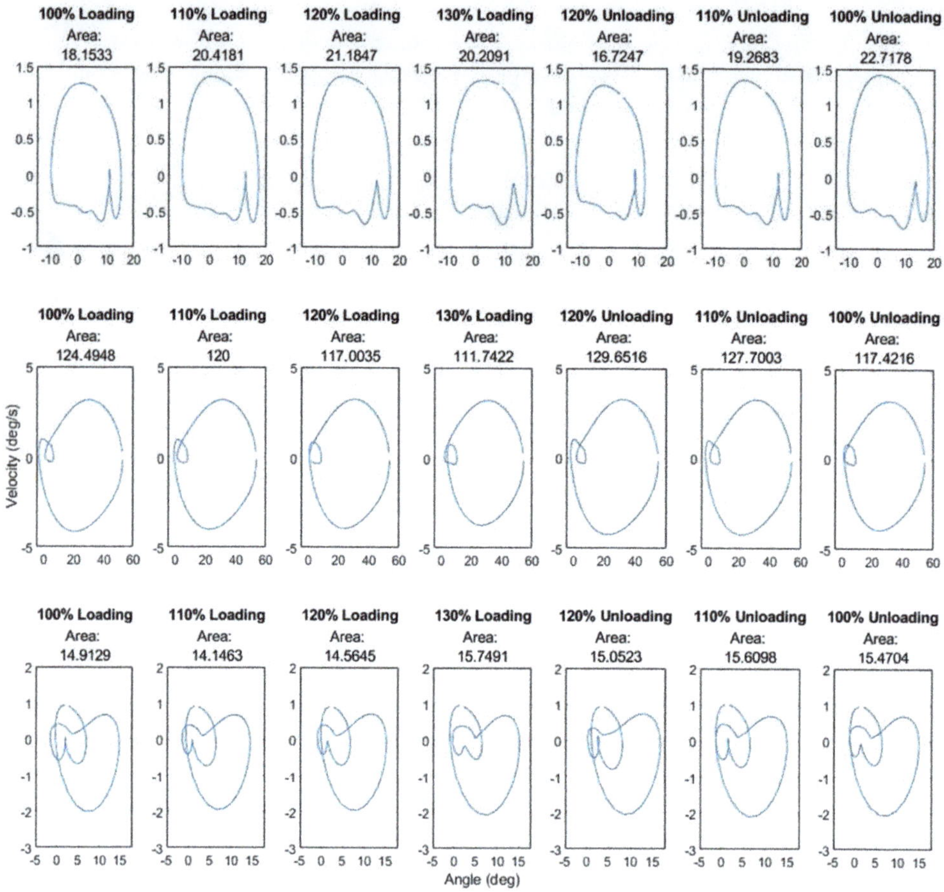

Figure 5. Phase portraits with calculated areas. The top row represents the hip phase portraits, the middle row is the knee phase portraits, and the bottom row is the ankle phase portraits. The area value represents the two-dimensional area—in pixels—occupied by each shape. These values can be considered a quantitative estimation of the range of available behaviors across each loading condition.

Table 2. Summed areas of phase portraits and percentage change.

Condition	% Load	Summed Areas of Hip, Knee and Ankle Phase Portraits	% Change from Previous Level of Loading
Loading	100%	157.561	
	110%	154.564	−2%
	120%	152.753	−1.1%
	130%	147.700	−3.3%
Unloading	120%	161.429	+9.2%
	110%	162.578	+0.7%
	100%	155.610	−4.2%

The results of this study have important implications for our understanding of the role that gravity plays in human locomotion. This study found that an increase in load can specifically affect both the knee and hip joints, as well as supporting the concept that loading and unloading are independent activities with specific responses. Even the same level of load, reached from higher or lower levels of weight, can elicit different responses.

In that way, increases in load appear to drive fewer hysteric changes than decreases (due to the relative availability of proprioceptive sensory information); as such, researchers should take care to ensure that their participants are responding to the correct level of load and should increase their load to the desired level, rather than unweight them.

There are also implications for using this system in long-duration spaceflight, where gravity may not be available to facilitate loading and gait. Indeed, the use of this system in a spacecraft could allow astronauts to maintain healthy levels of load for bone health and venous pumping, despite the absence of gravity, though research would clearly be needed to investigate this.

This study is not without limitations. There is a potential for this system to have influenced gait in some way, though no participant stated they felt the system interfered with their gait, arm swing, or ability to walk on the treadmill at any time. Similarly, participants were queried at all levels of load, and none expressed discomfort or fatigue with the system or any level of load. This study also did not compare its findings to more traditional loading studies examining the effects of external load on gait. As such, it is currently unclear how the effects seen in this study might compare with a heavy backpack or weight vest, for instance. However, while such a comparison was outside the scope of this study, future work should undoubtedly examine this. As the levels of load were not randomized, it is also possible that data could have been influenced by an order effect. This is a constraint of the system itself (individuals need to pass through any lower level of load in order to reach a specific level) which other potential users should be aware of.

This study was a novel investigation of an easily reproducible loading system that can increase the load of an individual, without the addition of external mass. Increased gravitational loading up to 130% of normal body weight can alter hip and knee kinematics but does not appear to affect the ankle joint nor does it appear to elicit changes in electromyographic variables. These findings suggest that the presence of external mass and alterations in limb inertias should be considered seriously as independent variables in future loading studies, and that weight and mass may need to be considered as separate effectors during locomotion. This study also found that the act of loading and unloading elicit distinct responses in the joints of the lower extremities, as well as that it may induce an adaptative after-effect.

Supplementary Materials: The following supporting information can be downloaded at: https://www.mdpi.com/article/10.3390/biomechanics5020031/s1, Table S1: EMG Peak by Muscle and Condition; Table S2: EMG RMS by Muscle and Condition; Table S3: EMG AUC by Muscle and Condition.

Author Contributions: All authors were involved in conceptualization, methodology, software, validation, formal analysis, investigation, resources, data curation, writing—original draft preparation, writing—review and editing visualization, supervision, project administration. All authors have read and agreed to the published version of the manuscript.

Funding: This research received no external funding.

Institutional Review Board Statement: This study was conducted in accordance with the Declaration of Helsinki and approved by the Institutional Review Board at the University of Houston (IRB#:00002971).

Informed Consent Statement: Informed consent was obtained from all participants involved in the study.

Data Availability Statement: Data can be made available upon request.

Conflicts of Interest: The authors declare no conflicts of interest.

References

1. Hirt, C.; Claessens, S.; Fecher, T.; Kuhn, M.; Pail, R.; Rexer, M. New ultrahigh-resolution picture of Earth's gravity field. *Geophys. Res. Lett.* **2013**, *40*, 4279–4283. [CrossRef]
2. Wolff, J. *Das Gesetz der Transformation der Knochen*; Hirchwild: Berlin, Germany, 1892.
3. Krasnov, I.B. Gravitational neuromorphology. *Adv. Space Biol. Med.* **1994**, *4*, 85–110. [PubMed]
4. Tairbekov, M.G. Kletka kak gravitatsionno-zavisimaia biomekhanicheskaia sistema [The cell as a gravity-dependent biomechanic system]. *Aviakosm. Ekolog. Med.* **2000**, *34*, 3–17. [PubMed]
5. Liew, B.; Morris, S.; Netto, K. The effect of backpack carriage on the biomechanics of walking: A systematic review and preliminary meta-analysis. *J. Appl. Biomech.* **2016**, *32*, 614–629. [CrossRef] [PubMed]
6. Arellano, C.J.; O'Connor, D.P.; Layne, C.; Kurz, M.J. The Independent Effect of Added Mass on the Stability of the Sagittal Plane Leg Kinematics during Steady-State Human Walking. *J. Exp. Biol.* **2009**, *212 Pt 12*, 1965–1970. [CrossRef] [PubMed]
7. Arellano, C.J.; Layne, C.S.; O'Connor, D.P.; Scott-Pandorf, M.; Kurz, M.J. Does Load Carrying Influence Sagittal Plane Locomotive Stability? *Med. Sci. Sports Exerc.* **2009**, *41*, 620–627. [CrossRef]
8. Boffey, D.; Harat, I.; Gepner, Y.; Frosti, C.L.; Funk, S.; Hoffman, J.R. The physiology and biomechanics of load carriage performance. *Mil. Med.* **2019**, *184*, e83–e90. [CrossRef]
9. Singh, T.; Koh, M. Effects of backpack load position on spatiotemporal parameters and trunk forward lean. *Gait Posture* **2009**, *29*, 49–53. [CrossRef]
10. Atwells, R.L.; Birrell, S.A.; Hooper, R.H.; Mansfield, N.J. Influence of carrying heavy loads on soldiers' posture, movements and gait. *Ergonomics* **2006**, *49*, 1527–1537. [CrossRef]
11. Polcyn, A.F.; Bensel, C.K.; Harman, E.A.; Obusek, J.P. *The Effects of Load Weight: A Summary Analysis of Maximal Performance, Physiological, and Biomechanical Results from Four Studies on Load–Carriage Systems*; United States Army Research Institute of Environmental Medicine Technical Report: Natick, MA, USA, 2001.
12. Haddox, A.G.; Hausselle, J.; Azoug, A. Changes in segmental mass and inertia during pregnancy: A musculoskeletal model of the pregnant woman. *Gait Posture* **2020**, *76*, 389–395. [CrossRef] [PubMed]
13. Jensen, R.K.; Nassas, G. Growth of segment principal moments of inertia between four and twenty years. *Med. Sci. Sports Exerc.* **1988**, *20*, 594–604. [CrossRef] [PubMed]
14. Villalba-Galea, C.A. Hysteresis in voltage-gated channels. *Channels* **2017**, *11*, 140–155. [CrossRef]
15. Villalba-Galea, C.A.; Chiem, A.T. Hysteretic Behavior in Voltage-Gated Channels. *Front. Pharmacol.* **2020**, *11*, 579596. [CrossRef] [PubMed]
16. Wei, J.Y.; Simon, J.; Randić, M.; Burgess, P.R. Joint angle signaling by muscle spindle receptors. *Brain Res.* **1986**, *370*, 108–118. [CrossRef]
17. Xiao, Y.-F.; Chandler, N.; Dobrzynski, H.; Richardson, E.S.; TenBroek, E.M.; Wilhelm, J.J.; Sharma, V.; Varghese, A.; Boyett, M.R.; Iaizzo, P.A.; et al. Hysteresis in human HCN4 channels: A crucial feature potentially affecting sinoatrial node pacemaking. *Sheng Li Xue Bao* **2010**, *62*, 1–13.
18. Kim, H.; Moon, J.-Y.; Mashour, G.A.; Lee, U. Mechanisms of hysteresis in human brain networks during transitions of consciousness and unconsciousness: Theoretical principles and empirical evidence. *PLoS Comput. Biol.* **2018**, *14*, e1006424. [CrossRef]
19. Sayal, A.; Sousa, T.; Duarte, J.V.; Costa, G.N.; Martins, R.; Castelo-Branco, M. Identification of competing neural mechanisms underlying positive and negative perceptual hysteresis in the human visual system. *Neuroimage* **2020**, *221*, 117153. [CrossRef] [PubMed]
20. Artz, N.J.; Adams, M.A.; Dolan, P. Sensorimotor function of the cervical spine in healthy volunteers. *Clin. Biomech.* **2015**, *30*, 260–268. [CrossRef]
21. Weiler, H.T.; Awiszus, F. Influence of hysteresis on joint position sense in the human knee joint. *Exp. Brain Res.* **2000**, *135*, 215–221. [CrossRef]
22. Abdolvahab, M.; Carello, C. Functional distance in human gait transition. *Acta Psychol.* **2015**, *161*, 170–176. [CrossRef]
23. Aoi, S.; Yamashita, T.; Tsuchiya, K. Hysteresis in the gait transition of a quadruped investigated using simple body mechanical and oscillator network models. *Phys. Rev. E—Stat. Nonlinear Soft Matter Phys.* **2011**, *83 Pt 1*, 061909. [CrossRef] [PubMed]
24. Aoi, S.; Katayama, D.; Fujiki, S.; Tomita, N.; Funato, T.; Yamashita, T.; Senda, K.; Tsuchiya, K. A stability-based mechanism for hysteresis in the walk-trot transition in quadruped locomotion. *J. R. Soc. Interface* **2013**, *10*, 20120908. [CrossRef] [PubMed] [PubMed Central]
25. Mohler, B.J.; Thompson, W.B.; Creem-Regehr, S.H.; Pick, H.L., Jr.; Warren, W.H., Jr. Visual flow influences gait transition speed and preferred walking speed. *Exp. Brain Res.* **2007**, *181*, 221–228. [CrossRef]
26. Thorstensson, A.; Roberthson, H. Adaptations to changing speed in human locomotion: Speed of transition between walking and running. *Acta Physiol. Scand.* **1987**, *131*, 211–214. [CrossRef] [PubMed]

27. Layne, C.S.; Malaya, C.A.; Young, D.R.; Suter, B.; Holder, J.L., Jr. Comparison of Treadmill Gait Between a Pediatric-Aged Individual with *SYNGAP1*-Related Intellectual Disability and a Fraternal Twin. *Front. Hum. Neurosci.* **2022**, *16*, 918918. [CrossRef] [PubMed]
28. Layne, C.S.; Malaya, C.A.; Ravindran, A.S.; John, I.; Francisco, G.E.; Contreras-Vidal, J.L. Distinct Kinematic and Neuromuscular Activation Strategies During Quiet Stance and in Response to Postural Perturbations in Healthy Individuals Fitted With and Without a Lower-Limb Exoskeleton. *Front. Hum. Neurosci.* **2022**, *16*, 942551. [CrossRef]
29. Winter, D.A.; Rau, G.; Kadefors, R.; Broman, H.; De Luca, C.J. *Units, Terms and Standards in the Reporting of EMG Research. A Report of the Ad Hoc Committee of the International Society of Electrophysiological Kinesiology*; Boston University: Boston, MA, USA, 1980.
30. Cifrek, M.; Medved, V.; Tonković, S.; Ostojić, S. Surface EMG based muscle fatigue evaluation in biomechanics. *Clin. Biomech.* **2009**, *24*, 327–340. [CrossRef]
31. Barton, P.M.; Hayes, K.C. Neck flexor muscle strength, efficiency, and relaxation times in normal subjects and subjects with unilateral neck pain and headache. *Arch. Phys. Med. Rehabil.* **1996**, *77*, 680–687. [CrossRef]
32. Carpentier, A.; Duchateau, J.; Hainaut, K. Motor unit behaviour and contractile changes during fatigue in the human first dorsal interosseus. *J. Physiol.* **2001**, *534 Pt 3*, 903–912. [CrossRef]
33. Linssen, W.H.J.P.; Stegeman, D.F.; Joosten, E.M.G.; Notermans, S.L.H.; van't Hof, M.A.; Binkhorst, R.A. Variability and interrelationships of surface EMG parameters during local muscle fatigue. *Muscle Nerve* **1993**, *16*, 849–856. [CrossRef]
34. van der Hoeven, J.H.; van Weerden, T.W.; Zwarts, M.J. Long-lasting supernormal conduction velocity after sustained maximal isometric contraction in human muscle. *Muscle Nerve* **1993**, *16*, 312–320. [CrossRef] [PubMed]
35. Bringoux, L.; Blouin, J.; Coyle, T.; Ruget, H.; Mouchnino, L. Effect of gravity-like torque on goal-directed arm movements in microgravity. *J. Neurophysiol.* **2012**, *107*, 2541–2548. [CrossRef] [PubMed]
36. Lackner, J.R.; DiZio, P. Gravitoinertial force level affects the appreciation of limb position during muscle vibration. *Brain Res.* **1992**, *592*, 175–180. [CrossRef]
37. Young, L.R.; Oman, C.M.; Merfeld, D.; Watt, D.; Roy, S.; DeLuca, C.; Balkwill, D.; Christie, J.; Groleau, N.; Jackson, D.K.; et al. Spatial orientation and posture during and following weightlessness: Human experiments on Spacelab Life Sciences 1. *J. Vestib. Res.* **1993**, *3*, 231–239. [CrossRef]
38. Mouchnino, L.; Cincera, M.; Fabre, J.C.; Assaiante, C.; Amblard, B.; Pedotti, A.; Massion, J. Is the regulation of the center of mass maintained during leg movement under microgravity conditions? *J. Neurophysiol.* **1996**, *76*, 1212–1223. [CrossRef] [PubMed]
39. Roll, R.; Gilhodes, J.C.; Roll, J.P.; Popov, K.; Charade, O.; Gurfinkel, V. Proprioceptive information processing in weightlessness. *Exp. Brain Res.* **1998**, *122*, 393–402. [CrossRef]
40. Mouchnino, L.; Lhomond, O.; Morant, C.; Chavet, P. Plantar Sole Unweighting Alters the Sensory Transmission to the Cortical Areas. *Front. Hum. Neurosci.* **2017**, *11*, 220. [CrossRef]
41. Saradjian, A.H.; Tremblay, L.; Perrier, J.; Blouin, J.; Mouchnino, L. Cortical facilitation of proprioceptive inputs related to gravitational balance constraints during step preparation. *J. Neurophysiol.* **2013**, *110*, 397–407. [CrossRef]
42. Thiel, S.D.; Bitzer, S.; Nierhaus, T.; Kalberlah, C.; Preusser, S.; Neumann, J.; Nikulin, V.V.; van der Meer, E.; Villringer, A.; Pleger, B. Hysteresis as an implicit prior in tactile spatial decision making. *PLoS ONE* **2014**, *9*, e89802. [CrossRef] [PubMed] [PubMed Central]
43. Kostyukov, A.I.; Cherkassky, V.L. Interaction of the movement-dependent, extrafusal and fusimotor after-effects in the firing of the primary spindle endings. *Neuroscience* **1997**, *76*, 1257–1266. [CrossRef]
44. Kozlovskaya, I.B.; Sayenko, I.V.; Sayenko, D.G.; Miller, T.F.; Khusnutdinova, D.R.; Melnik, K.A. Role of support afferentation in control of the tonic muscle activity. *Acta Astronaut.* **2007**, *60*, 285–294. [CrossRef]
45. Shenkman, B.S.; Kozlovskaya, I.B. Cellular responses of human postural muscle to dry immersion. *Front. Physiol.* **2019**, *10*, 187. [CrossRef] [PubMed]
46. Shenkman, B.S.; Grigoriev, A.I.; Kozlovskaya, I.B. Gravity mechanisms in tonic motor system. Neurophysiological and Muscle Aspects. *Hum. Physiol.* **2017**, *43*, 578–590. [CrossRef]
47. Saveko, A.; Brykov, V.; Kitov, V.; Shpakov, A.; Tomilovskaya, E. Adaptation in Gait to Lunar and Martian Gravity Unloading During Long-Term Isolation in the Ground-Based Space Station Model. *Front. Hum. Neurosci.* **2022**, *15*, 742664. [CrossRef]
48. Shpakov, A.V.; Artamonov Voronov, A.A.; Melnik, K.A. Influence of immersion hypokinesia on kinematic and electromyographic characteristics of human locomotion. *Aviakosm. Ekolog. Med.* **2008**, *5*, 24–29.
49. Litvinova, K.S.; Vikhlyantsev, I.M.; Kozlovskaya, I.B.; Podlubnaya, Z.A.; Shenkman, B.S. Effects of artificial support stimulation on fiber and molecular characteristics of soleus muscle in men exposed to 7-day dry immersion. *J. Gravit. Physiol.* **2004**, *11*, 131–132.
50. Stergiou, N. *Innovative Analyses of Human Movement*; Human Kinetics: Champaign, IL, USA, 2004.

51. Manzano, M.; Serrancolí, G. A factorization-based algorithm to predict EMG data using only kinematics information. *Int. J. Numer. Methods Biomed. Eng.* **2021**, *37*, e3463. [CrossRef]
52. Mauntel, T.C.; Cram, T.R.; Frank, B.S.; Begalle, R.L.; Norcross, M.F.; Blackburn, J.T.; Padua, D.A. Kinematic and neuromuscular relationships between lower extremity clinical movement assessments. *Sports Biomech.* **2018**, *17*, 273–284. [CrossRef]

biomechanics

MDPI

Article

Curve Analysis of Foot Coupling Kinematics in Runners with Plantar Heel Pain During Running Gait

Hanieh Pazhooman [1], Robin L. Pomeroy [1], Mohammed S. Alamri [1,2], Zahra Mollaei [1] and Stephen C. Cobb [1,*]

[1] Foot & Ankle Biomechanics Laboratory, University of Wisconsin–Milwaukee, University Services and Research Building, 115 E Reindl Way, Glendale, WI 53212, USA; pazhoom2@uwm.edu (H.P.); pomeroyro@gmail.com (R.L.P.); msalamri@uwm.edu (M.S.A.); zmollaei@uwm.edu (Z.M.)

[2] Biomechanics Laboratory, Department of Biomechanics and Motor Behavior, King Saud University, Riyadh 4545, Saudi Arabia

* Correspondence: cobbsc@uwm.edu

Abstract: (1) Background: Plantar heel pain (PHP), a common overuse foot injury, significantly impacts runners. While the mechanical role of the plantar fascia during gait is established, its effect on foot function during running, particularly foot joint coupling, remains unclear. This study investigated foot joint coupling during running in runners with and without PHP using statistical parametric mapping (SPM). (2) Methods: Thirteen uninjured runners (seven m, six f; age = 30.5 ± 5.9 years; BMI = 23.5 ± 3.0 kg/m^2) and thirteen runners with PHP (six m, seven f; age = 29.0 ± 8.0 years; BMI = 23.1 ± 2.0 kg/m^2) performed running trials at 4.0 m/s. A seven-segment foot model that defined six functional articulations (rearfoot, medial and lateral midfoot, medial and lateral forefoot, first metatarsophalangeal) was used to quantify foot kinematics, vector coding was used to calculate joint coupling between adjacent foot segments, and SPM was used to analyze joint stance phase coupling angles. (3) Results: There were statistically significant differences in rearfoot frontal plane–medial midfoot frontal plane joint coupling between runners with and without PHP from 69 to 70% stance phase (mean difference = 39.41°) and at 76% stance (mean difference = 47.89°). The differences were indicative of greater medial midfoot eversion rotation relative to rearfoot complex inversion in the PHP group. (4) Conclusions: The difference in the rearfoot complex and medial midfoot frontal plane coupling occurred during the propulsion phase of the running stance when the foot should be transitioning to a more supinated position, which may reflect compromised supination due to plantar fascia degeneration.

Keywords: plantar fasciitis; foot joint coordination; multi-segment foot; overuse injury

1. Introduction

Plantar heel pain involving the plantar fascia (plantar fasciitis, plantar fasciosis) is one of the most common overuse foot injuries in the general population, affecting both active and sedentary individuals [1]. With respect to the prevalence in active individuals, a prospective study of lower extremity injuries in runners found that plantar heel pain was the most common injury, affecting 31.3% of runners over the five-year study period [2]. Further, a systematic review of running-related musculoskeletal injuries reported plantar heel pain as the third most common injury after medial tibial stress syndrome and Achilles tendinopathy [3]. In addition to the high prevalence, plantar heel pain is also associated with significant economic and personal burden. In the United States, 65% of patients with plantar heel pain seek medical attention multiple times [4], and the average annual treatment cost associated with plantar heel pain is USD 284 million [5]. Moreover, studies have shown that plantar heel pain is associated with diminished foot-specific health and overall health-related quality of life [6]. To reduce the negative impacts associated with plantar heel pain, it is crucial to identify and address the pathoetiological factors associated with the injury.

In general, it is accepted that the mechanical overload of the plantar fascia during gait and/or long periods of weightbearing leads to microtrauma and subsequent degeneration [7]. During gait, the plantar fascia contributes as a passive stabilizer to control foot pronation in the early stance and to facilitate foot supination in the mid-late stance [8–14]. Prior to initial contact, the plantar fascia is preloaded to facilitate control of the foot pronation that occurs following initial contact [12]. During the early stance phase, tension in the plantar fascia effectively secures the cuboid within the confines of the calcaneus, thereby establishing stability within the midtarsal complex of the foot (calcaneocuboid locking mechanism) [8,11]. As the foot enters into midstance and the orientation of the midtarsal shafts shifts from a horizontal to vertical alignment, the plantar fascia assumes a tie-rod role, effectively distributing compressive loads across the tarsal and metatarsal bones and enhancing stability within the central midfoot structures (truss mechanism) [8]. Finally, during the terminal phase of the gait, tension in the plantar fascia activates the windlass mechanism, which facilitates foot supination [8,13]. While the function of the plantar fascia during gait is well established, the effect of plantar heel pain on foot function during running gait is currently poorly understood. The current clinical belief is that the repetitive straining of the plantar fascia during running causes progressive chronic degeneration that is eventually experienced as heel pain [15]. In addition to producing pain, degeneration [7,16] is also likely to cause changes to the mechanical properties of the plantar fascia and compromised function during running gait. The compromised function during gait may contribute to the development and/or persistence of the plantar heel pain. To date, only two studies have investigated the effect of plantar heel pain on foot function during running gait [17,18]. Pazhooman et al. [17] reported differences in the stance phase range of motion in runners with plantar heel pain, consistent with a decreased ability to control foot pronation in the early–mid stance and contribute to foot supination in the mid-late stance. Wiegand et al. [18] reported significant differences between runners with plantar heel pain and runners that had recovered from plantar heel pain, but not between runners with plantar heel pain and uninjured runners.

In addition to changes in the kinematics at individual joints, the coupling between adjacent joints has also been theorized to be important to normal lower extremity function during gait. Changes in the normal/typical coupling between adjacent joints are believed to cause the abnormal loading of the bones and soft tissues in and around the affected joints, which may lead to initial injury and/or injury persistence [19]. Thus, abnormal joint coupling between adjacent joints of the foot may be a cause of repetitive strain, leading to plantar heel pain and/or may contribute to the persistence of plantar heel pain. The only study conducted to date found that individuals with plantar heel pain demonstrated significantly fewer frontal plane anti-phase movements between the rearfoot and medial forefoot during walking gait than uninjured participants [20]. While this study has further improved the understanding of the effect of plantar heel pain on foot function during walking gait, the results may not be generalizable to running gait [21]. Furthermore, this study investigated discrete variables at predetermined instances/periods during gait. While discrete analyses are important, they may not capture other relevant changes throughout the stance phase. Statistical parametric mapping (SPM) is an approach that enables the examination of joint kinematics across an entire time series [22].

To date, the effect of plantar heel pain on foot joint coupling during running gait has not been investigated using curve analysis techniques. Therefore, the purpose of this study was to employ SPM methodology to investigate foot joint coupling during running gait in runners with and without plantar heel pain. We hypothesized that the degenerative changes in the plantar fascia that occur with plantar heel pain would change the foot joint coupling angles of individuals with plantar heel pain during the stance phase of running gait.

2. Materials and Methods

A total of 13 uninjured runners (7 m, 6 f) (uninjured) and 13 runners with plantar heel pain associated with the pathology of plantar fascia (PHP) (6 m, 7 f) participated in the study. All runners were between the ages of 18 and 45 years old, and all were informed of the study procedures and provided written consent approved by the Institutional Review Board. The PHP runners had common clinical symptoms of plantar heel pain associated with pathology of the plantar fascia for a minimum of six weeks, were consistently running at least 10 miles per week at the time of this study [23], and had a body mass index (BMI) of <30 kg/m^2 [24]. The clinical symptoms indicative of plantar heel pain consisted of point tenderness localized to the medial calcaneal tubercle, the medial region of the proximal segment of the plantar fascia, or along the length of the plantar fascia [25]. In addition, runners must have reported plantar heel pain during the first few steps of weight-bearing, following prolonged periods of inactivity that gradually diminished throughout the day with normal walking [26–28]. The presence of plantar heel pain due to pathology of the plantar fascia was confirmed, and other potential sources of heel pain (e.g., heel pad syndrome, tibialis posterior tendinopathy) were ruled out during a brief screening exam performed by an allied health professional. Runners in the uninjured group did not have an history of plantar heel pain and were age (±5 y), sex, and mileage (10–20, 20–30, 30+ miles per week)-matched with the PHP runners. Exclusion criteria for all participants were as follows: current lower extremity injury, besides plantar heel pain in the PHP group; pregnancy; a history of lower extremity surgery on the injured side; inflammatory or connective tissue disease; diagnosed foot deformity (e.g., hallux valgus); a systemic neurologic disorder that would predispose an individual to heel pain and/or muscle weakness; and diabetic neuropathy [29].

To determine if the current investigation would have sufficient power to detect significant differences between groups, a power analysis using data from the Wiegand et al. [18] study that investigated the effect of plantar heel pain on foot function during running was used. Results of the power analysis (G Power 3.1) indicated that a sample size of 26 participants (13 participants per group) would be sufficient to identify a large effect size (Cohen d = 0.9), at an alpha level of 0.05, and a power 0.8.

Foot kinematics were quantified using a seven-segment foot model that defines six foot segments and a shank segment (Figure 1) [30]. The multi-segment foot model is fully described in our previous studies [17,30]. Following the application of the retroreflective markers associated with the foot model, a static trial was performed to identify additional anatomical landmarks relative to the appropriate technical marker cluster. Participants then completed 10 successful running trials at 4.0 m/s (±10%) along a runway wearing the same style sandals (Maui and Sons, Pacific Palisades, CA, USA) [31]. The sandals were selected because they enabled the placement of markers directly on the participant's skin and allowed participants to utilize their preferred foot strike pattern during running. Running gait speed was monitored using an opto-electronic timing device. A successful trial was defined as a trial within the fixed speed range where initial contact and toe off occurred on the force plate. Three-dimensional (3D) marker positions were captured at 200 Hz with a 10-camera Eagle system (Motion Analysis Inc, Santa Rosa, CA, USA). The stance phase initial contact and toe-off events (10 N threshold) were identified by a force plate (Advanced Mechanical Technology, Inc., Watertown, MA, USA) sampling at 1000 Hz mounted near the center of the runway.

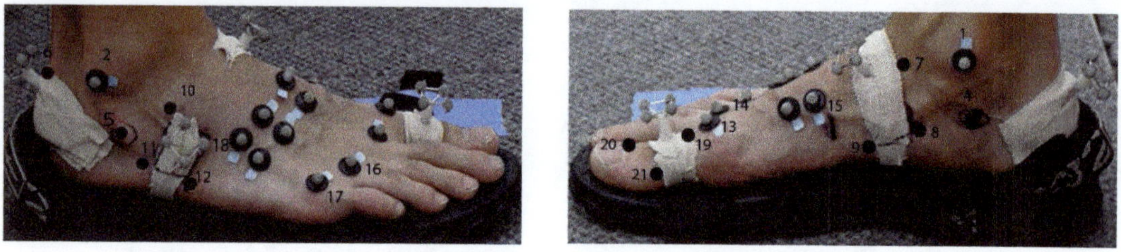

Figure 1. Anatomical landmarks: Shank [medial malleolus (1), lateral malleolus (2), tibial tuberosity (3, not pictured)]; **calcaneus** [sustentaculum tali (4), peroneal tubercle (5), dorsal proximal calcaneus (6)]; **navicular** [dorsal proximal (7), plantar proximal (8), plantar distal (9)]; **cuboid** [dorsal proximal (10), plantar proximal (11), plantar distal (12)]; **medial rays** [1st metatarsal head (13), 2nd metatarsal head (14), 1st metatarsal base (15)]; **lateral rays** [4th metatarsal head (16), 5th metatarsal head (17), base of 5th metatarsal (18)]; **hallux** [base pf proximal phalanx (19), head of distal phalanx (20), medial aspect (21)]. **Functional articulations: rearfoot complex** (shank–calcaneus); **medial midfoot** (calcaneus–navicular); **lateral midfoot** (calcaneus–cuboid); **medial forefoot** (navicular–medial rays); **lateral forefoot** (cuboid–lateral rays); **first metatarsophalangeal** (medial rays–hallux).

Following the static trial and running trials, 3D marker positions were reconstructed using Cortex software (v. 5.3, Motion Analysis Inc, Santa Rosa, CA, USA). A custom-written software program (MATLAB v. 7.6.0, The MathWorks Inc., Natick, MA, USA) was then used to low-pass filter (4th order zero-lag Butterworth filter) the kinematic (12 Hz cutoff) and force plate (50 Hz cutoff) data and perform rigid body transformation procedures. Joint angles computed during the static trial were used as offset angles for the gait trials. For the running trials, the first five trials with no obvious erroneous marker displacement were selected for the stance phase coupling analysis. Stance phase joint coupling angles between adjacent functional articulations of interest (Table 1) were calculated using the vector coding method proposed by Heiderscheit et al. [32] (Equation (1)).

$$\theta_i=\tan^{-1}\left(\frac{y_{i+1}-y_i}{x_{i+1}-x_i}\right) \quad (1)$$

where θ_i is the joint coupling angle between 0° and 360°, i is the ith frame of the stance phase, and x and y are the angular positions of the proximal and distal functional articulation, respectively. The angle calculated is the orientation to the right horizontal of the resultant vector between two adjacent data points of the proximal functional articulation (e.g., the shank for the rearfoot complex) and two adjacent data points of the distal segment (e.g., the calcaneus for the rearfoot complex) during the stance phase [32]. The coupling angles were interpreted as defined by Hamill et al. [33].

Table 1. Joint couples investigated.

RC	RC-MMF	RC-LMF	MMF-MFF	LMF-LFF	MFF-1MTP
RC_{Fron}-RC_{Tran}	RC_{Sag}-MMF_{Sag}	RC_{Sag}-LMF_{Sag}	MMF_{Fron}-MFF_{Sag}	LMF_{Fron}-LFF_{Sag}	MFF_{Sag}-$1MTP_{Sag}$
	RC_{Fron}-MMF_{Fron}	RC_{Fron}-LMF_{Fron}	MMF_{Fron}-MFF_{Fron}	LMF_{Fron}-LFF_{Fron}	
	RC_{Tran}-MMF_{Tran}	RC_{Tran}-LMF_{Tran}			

RC = rearfoot Complex; MMF = medial midfoot functional articulation; LMF = lateral midfoot functional articulation; MFF = medial forefoot functional articulation; LFF = lateral forefoot functional articulation; 1MTP = first metatarsophalangeal functional articulation; Fron = frontal plane; Sag = sagittal plane; Trans = transverse plane.

- 0° or 180° indicates rotation primarily of the proximal joint with the distal joint fixed.

- 90° or 270° indicates rotation primarily of the distal joint with the proximal joint fixed.
- 45° or 225° reflects equal relative rotation of joints in the same direction.
- 135° or 315° indicates equal relative rotation of the two joints in opposite directions.

Finally, the spm1d package for one-dimensional SPM was used in MATLAB to investigate the joint coupling time series using circular statistics. To assess the normality of the data, all the comparisons were completed using the parametric (spm1d.stats.ttest2) and nonparametric (spm1d.stats.nonparam) scripts. If the parametric and nonparametric results did not differ qualitatively, the assumption of normality was considered reasonable, and the results of the parametric t-test were used. The independent variable in each test was the group (PHP, uninjured). The dependent variables were the joint couples of interest during stances. Significance for all statistical tests was defined as $\alpha = 0.05$.

3. Results

Descriptive data are presented in Table 2. A previously published statistical analysis of the descriptive data for this cohort of participants reported no significant age, BMI, or running speed group differences [17]. A rearfoot strike pattern was the preferred strike pattern of all the runners. The strike pattern was observed during the gait trials and was confirmed following the trials by the presence of rearfoot complex plantar flexion in the weight acceptance period following initial contact.

Table 2. Participant descriptive data, mean (SD).

Variable	Uninjured	PHP
Age (years)	30.5 (5.9)	29.0 (8.0)
BMI (kg/m^2)	23.5 (3.0)	23.1 (2.0)
Running speed (m/s)	4.1 (0.2)	4.0 (0.3)

PHP: Plantar heel pain.

None of parametric and nonparametric SPM results differed qualitatively, so the parametric t-test was selected for all the comparisons. The SPM analyses between runners with and without plantar heel pain revealed a significant difference in RC_{Fron}-MMF_{Fron} joint coupling between the PHP and uninjured groups from 69 to 70% (mean difference = 39.41°, $p = 0.042$, 95% CI [36.2°, 42.6°]) stance phase and at 76% (mean difference = 47.89°, $p = 0.049$, 95% CI [44.7°, 51.1°]) of the running gait stance phase (Figure 2). During 69–70% stance, the mean coupling angles of the PHP and uninjured groups were 291.6 ± 1.94° and 331.0 ± 1.9°, respectively. At 76% stance, the coupling angles of the PHP and uninjured groups were 264.2° and 312.1°, respectively. The results of the SPM analyses that were not statistically significant can be found in the Supplementary Materials (Figures S1–S11).

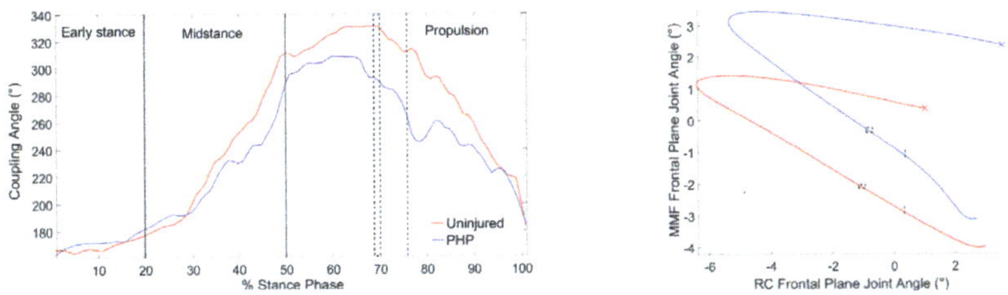

Figure 2. Rearfoot complex frontal plane and medial midfoot frontal plane vector coding joint coupling time series (**left** figure) and angle–angle time series (**right** figure). The red line is the mean of the uninjured runners, the blue line is the mean of the runners with plantar heel pain. The vertical bars on the left figure identify the subphase events. The dashed boxes on both figures indicate the areas of significant group differences. The x's in the right figure identify initial contact of stance phase.

4. Discussion

The purpose of the current study was to use SPM methodology to investigate foot joint coupling during the stance phase of running gait in runners with and without plantar heel pain. Our hypothesis that the foot coupling angles would be different in runners with plantar heel pain compared to uninjured runners during running was supported for the rearfoot complex frontal plane and medial midfoot frontal plane coupling angle between 69 and 70% stance phase and at 76% stance. Both differences occurred during the propulsion subphase (50–100%) of the running stance, when the foot should be transitioning to a more supinated position.

From 69 to 70% stance, the mean coupling angle of the PHP group was 291.6 ± 1.94°, which is between "relative equal rotation of the two joints in opposite directions (315°)" and "rotation primarily of the distal joint with the proximal joint fixed (270°)", indicating the rearfoot complex and medial midfoot were rotating in opposite directions with the medial midfoot rotating more than the rearfoot complex. The angle–angle plot in Figure 2 indicates the rearfoot complex was in an everted position and inverting, while the medial midfoot was in an inverted position and everting. During the same period, the mean coupling angle of the uninjured group was 331.0 ± 1.9°, which is between "rotation primarily of the proximal joint with the distal joint fixed (0°)" and "relative equal rotation of the two joints in opposite directions (315°)", indicating that the rearfoot complex and medial midfoot were rotating in opposite directions, but with the rearfoot complex rotating more than the medial midfoot. The angle–angle plot in Figure 2 indicates the rearfoot complex was in an everted position and inverting, while the medial midfoot was in an everted position and everting.

At 76% stance, the PHP group coupling angle of 264.2° is very close to "rotation primarily of the distal joint with the proximal joint fixed (270°)", indicating the medial midfoot was the primary joint rotating and the rearfoot was relatively fixed. Finally, the coupling angle of the uninjured groups at 76% stance of 312.1° is very close to "equal relative motion of the two joints in opposite directions (315°)", indicating the near equal relative rotation of the rearfoot and medial midfoot in opposite directions. The difference in the rearfoot complex and medial midfoot frontal plane coupling occurred during the propulsion phase of the running stance phase, when the foot should be transitioning to a more supinated position. The greater midfoot eversion rotation relative to rearfoot inversion rotation in the PHP group could be an indicator of insufficient supination due to the degenerative changes in the plantar fascia.

Because this is the first study to utilize SPM to examine foot joint coupling in this population during running, the results cannot be directly compared to previous plantar heel pain gait studies. The only previous studies investigating the effect of plantar heel pain on running gait included a discrete analysis of the stance phase range of motion using the same cohort of runners as the current study [17], and the study by Wiegand et al. [18]. In the Pazhooman et al. [17] study, the runners with PHP demonstrated increased lateral midfoot eversion ROM during the early stance, which may be consistent with a decreased ability to control foot pronation in the early–mid stance. Interestingly, the difference identified between the groups using SPM to investigate coupling between adjacent joints was also in the frontal plane; however, it did not involve the lateral midfoot and occurred in the propulsion phase vs. early stance. These differences suggest that discrete single-joint ROM analysis and SPM analysis may detect different characteristics of plantar heel pain. A strength of SPM analysis is that it can detect significant differences anywhere in the time series versus at predetermined events (e.g., peaks) or within predetermined periods (e.g., early stance). Wiegand et al. [18] did not report any significant differences in sagittal or frontal plane forefoot, midfoot, or ankle peak angles or ROM during stances between runners with plantar heel pain and uninjured runners. In addition to the difference in the statistical analysis techniques utilized, the inclusion criteria (symptoms within the past two weeks vs. for a minimum of six weeks), footwear (participant's personal running shoe vs. standardized sandal), and marker placement (on participant's running shoe vs. on participant's skin) may have contributed to the inconsistent results between the studies.

The only previous study to investigate joint coupling in individuals with plantar heel pain was a discrete analysis that reported a greater number of anti-phase movements between the rearfoot complex and medial forefoot joints in the frontal plane in uninjured individuals compared to individuals with PHP during the stance phase of walking gait [20]. The results of the Chang et al. [20] study and those of the current study cannot be compared due to difference in the study methodologies (discrete vs. continuous, walking gait vs. running gait). With respect to gait type, it is possible that PHP may affect foot function differently during walking compared to running due to the increased forces and/or decreased time during stances for the foot to accomplish its various functions during running. The likelihood that the results of the effect on running may not be generalizable to walking, and vice-versa, is supported by a previous study that investigated the effect of medial tibial stress syndrome (MTSS) on foot function during walking and running [34]. This study found that the effect of MTSS on hindfoot and forefoot motion, compared to an uninjured group, differed between walking and running gait.

Before drawing conclusions regarding this study, several limitations should be considered. First, to minimize variability due to running speed, all runners ran at a fixed speed (4 m/s $\pm$ 10%). Although the running speed range is common for healthy adults, it may not have represented every runner's preferred speed, which could influence their foot kinematics. Second, to accommodate multi-segment foot marker placement and eliminate variability that could have been caused by differing footwear, all participants wore the same style sandal (Maui and Sons, Pacific Palisades, CA, USA). As a result, the foot kinematics in the sandal may not be generalizable to the foot coupling kinematics that occur in traditional running shoes. It will be important for future studies to investigate if varying degrees of footwear support (e.g., minimalist vs. motion control) influence foot coupling during running gait in patients with PHP. Third, this study only included young adults, limiting the generalizability of the findings to other age groups. Fourth, all the runners in the current study were rearfoot strikers. Thus, the results may not be generalizable to runners with forefoot strike patterns [35,36]. Fifth, to have sufficient statistical power for the SPM analyses, the current analysis combined male and female runners in the PHP and uninjured groups. However, recent studies have identified sex differences in foot kinematics in both healthy [37] and injured [17,38] individuals. Thus, it will be important for future studies to explore a potential interaction between sex and group (uninjured vs. PHP) on foot joint coupling during running gait. Sixth, the current study was sufficiently powered to identify large effect sizes but may not have been able to identify moderate effect sizes, which may also be clinically relevant. Lastly, the current study was a retrospective study; therefore, whether the observed differences in foot function are a cause or a consequence of plantar heel pain cannot be determined.

5. Conclusions

This is the first study to utilize SPM to investigate foot joint coupling kinematics during running in runners with plantar heel pain. Our results suggest that plantar heel pain significantly affects the frontal plane coupling between the rearfoot and medial midfoot during the propulsion phase of running gait. The difference may contribute to the development and/or persistence of plantar heel pain in runners. While future prospective studies are needed to determine whether foot joint coupling differences are a cause or consequence of plantar heel pain, the current findings could also serve as an additional outcome measure in future intervention studies.

Supplementary Materials: The following supporting information can be downloaded at: https://www.mdpi.com/article/10.3390/biomechanics5020034/s1, Figures S1–S11.

Author Contributions: Conceptualization, R.L.P., S.C.C. and H.P.; methodology, R.L.P., S.C.C., H.P., M.S.A. and Z.M.; software, H.P., S.C.C. and R.L.P.; validation, H.P. and S.C.C.; formal analysis, H.P. and S.C.C.; investigation, R.L.P.; resources, S.C.C.; data curation, H.P., S.C.C. and R.L.P.; writing—original draft preparation, H.P. and S.C.C.; writing—review and editing, H.P., R.L.P., M.S.A., Z.M. and S.C.C.; visualization, H.P. and S.C.C.; supervision, S.C.C.; project administration, S.C.C.; funding acquisition, R.L.P. All authors have read and agreed to the published version of the manuscript.

Funding: This work was supported by UW-Milwaukee, College of Health Sciences, Student Research Grant Award, and the American Society of Biomechanics Graduate Student Grant-in-Aid.

Institutional Review Board Statement: This study was approved by the Institutional Review Board.

Informed Consent Statement: Written informed consent was obtained from all participants of this study.

Data Availability Statement: Available upon reasonable request.

Conflicts of Interest: The authors declare no conflicts of interest.

References

1. Gill, T.K.; Menz, H.B.; Landorf, K.B.; Arnold, J.B.; Taylor, A.W.; Hill, C.L. Identification of Clusters of Foot Pain Location in a Community Sample. *Arthritis Care Res.* **2017**, *69*, 1903–1908. [CrossRef]
2. Di Caprio, F.; Buda, R.; Mosca, M.; Calabro, A.; Giannini, S. Foot and Lower Limb Diseases in Runners: Assessment of Risk Factors. *J. Sports Sci. Med.* **2010**, *9*, 587–596. [PubMed]
3. Lopes, A.D.; Hespanhol Júnior, L.C.; Yeung, S.S.; Costa, L.O. What are the Main Running-Related Musculoskeletal Injuries? A Systematic Review. *Sports Med.* **2012**, *42*, 891–905. [CrossRef] [PubMed]
4. Riddle, D.L.; Schappert, S.M. Volume of Ambulatory Care Visits and Patterns of Care for Patients Diagnosed with Plantar Fasciitis: A National Study of Medical Doctors. *Foot Ankle Int./Am. Orthop. Foot Ankle Soc. Swiss Foot Ankle Soc.* **2004**, *25*, 303–310. [CrossRef]
5. Tong, K.B.; Furia, J. Economic Burden of Plantar Fasciitis Treatment in the United States. *Am. J. Orthop.* **2010**, *39*, 227–231.
6. Irving, D.B.; Cook, J.L.; Young, M.A.; Menz, H.B. Impact of Chronic Plantar Heel Pain on Health-Related Quality of Life. *J. Am. Podiatr. Med. Assoc.* **2008**, *98*, 283–289. [CrossRef] [PubMed]
7. Tountas, A.A.; Fornasier, V.L. Operative Treatment of Subcalcaneal Pain. *Clin. Orthop. Relat. Res.* **1996**, *332*, 170–178. [CrossRef]
8. Aquino, A.; Payne, C. Function of the Plantar Fascia. *Foot* **1999**, *9*, 73–78. [CrossRef]
9. Hicks, J.H. The Mechanics of the Foot. I. The Joints. *J. Anat.* **1953**, *87*, 345–357.
10. Dananberg, H.J. Gait Style as an Etiology of Chronic Postural Pain: Part I. Functional Hallux Limitus. *J. Am. Podiatr. Med. Assoc.* **1993**, *83*, 433–441. [CrossRef]
11. Bojsen-Moller, F. Calcaneocuboid Joint and Stability of the Longitudinal Arch of the Foot at High and Low Gear Push Off. *J. Anat.* **1979**, *129*, 165–176. [PubMed]
12. Caravaggi, P.; Pataky, T.; Goulermas, J.Y.; Savage, R.; Crompton, R. A Dynamic Model of the Windlass Mechanism of the Foot: Evidence for Early Stance Phase Preloading of the Plantar Aponeurosis. *J. Exp. Biol.* **2009**, *212*, 2491–2499. [CrossRef] [PubMed]
13. Welte, L.; Kelly, L.A.; Kessler, S.E.; Lieberman, D.E.; D'Andrea, S.E.; Lichtwark, G.A.; Rainbow, M.J. The Extensibility of the Plantar Fascia Influences the Windlass Mechanism during Human Running. *Proc. Biol. Sci.* **2021**, *288*, 20202095. [CrossRef]
14. Schleip, R.; Gabbiani, G.; Wilke, J.; Naylor, I.; Hinz, B.; Zorn, A.; Jäger, H.; Breul, R.; Schreiner, S.; Klingler, W. Fascia Is Able to Actively Contract and May Thereby Influence Musculoskeletal Dynamics: A Histochemical and Mechanographic Investigation. *Front. Physiol.* **2019**, *10*, 336. [CrossRef]
15. Buchanan, B.K.; Sina, R.E.; Kushner, D. Plantar Fasciitis. In *StatPearls*; StatPearls Publishing: Treasure Island, FL, USA, 2024.
16. Leach, R.E.; Seavey, M.S.; Salter, D.K. Results of Surgery in Athletes with Plantar Fasciitis. *Foot Ankle* **1986**, *7*, 156–161. [CrossRef]
17. Pazhooman, H.; Alamri, M.S.; Pomeroy, R.L.; Cobb, S.C. Foot Kinematics in Runners with Plantar Heel Pain during Running Gait. *Gait Posture* **2023**, *104*, 15–21. [CrossRef]

18. Wiegand, K.; Tandy, R.; Freedman Silvernail, J. Plantar Fasciitis Injury Status Influences Foot Mechanics during Running. *Clin. Biomech.* **2022**, *97*, 105712. [CrossRef] [PubMed]
19. DeLeo, A.T.; Dierks, T.A.; Ferber, R.; Davis, I.S. Lower Extremity Joint Coupling during Running: A Current Update. *Clin. Biomech.* **2004**, *19*, 983–991. [CrossRef]
20. Chang, R.; van Emmerik, R.; Hamill, J. Chronic Plantar Fasciitis Reduces Rearfoot to Medial-Forefoot Anti-Phase Coordination. *Clin. Biomech.* **2021**, *88*, 105439. [CrossRef]
21. Pohl, M.B.; Messenger, N.; Buckley, J.G. Forefoot, Rearfoot and Shank Coupling: Effect of Variations in Speed and Mode of Gait. *Gait Posture* **2007**, *25*, 295–302. [CrossRef]
22. Pataky, T.C. Generalized N-Dimensional Biomechanical Field Analysis Using Statistical Parametric Mapping. *J. Biomech.* **2010**, *43*, 1976–1982. [CrossRef] [PubMed]
23. Messier, S.P.; Pittala, K.A. Etiologic Factors Associated with Selected Running Injuries. *Med. Sci. Sports Exerc.* **1988**, *20*, 501–505. [CrossRef] [PubMed]
24. Riddle, D.L.; Pulisic, M.; Pidcoe, P.; Johnson, R.E. Risk Factors for Plantar Fasciitis: A Matched Case-Control Study. *J. Bone Jt. Surg. Am.* **2003**, *85*, 872–877. [CrossRef]
25. Cornwall, M.W.; McPoil, T.G. Plantar Fasciitis: Etiology and Treatment. *J. Orthop. Sports Phys. Ther.* **1999**, *29*, 756–760. [CrossRef]
26. Karabay, N.; Toros, T.; Hurel, C. Ultrasonographic Evaluation in Plantar Fasciitis. *J. Foot Ankle Surg.* **2007**, *46*, 442–446. [CrossRef]
27. Allen, R.H.; Gross, M.T. Toe Flexors Strength and Passive Extension Range of Motion of the First Metatarsophalangeal Joint in Individuals with Plantar Fasciitis. *J. Orthop. Sports Phys. Ther.* **2003**, *33*, 468–478. [CrossRef]
28. Cardinal, E.; Chhem, R.K.; Beauregard, C.G.; Aubin, B.; Pelletier, M. Plantar Fasciitis: Sonographic Evaluation. *Radiology* **1996**, *201*, 257–259. [CrossRef]
29. Wearing, S.C.; Smeathers, J.E.; Urry, S.R.; Hennig, E.M.; Hills, A.P. The Pathomechanics of Plantar Fasciitis. *Sports Med.* **2006**, *36*, 585–611. [CrossRef]
30. Cobb, S.C.; Joshi, M.N.; Pomeroy, R.L. Reliability of a Seven-Segment Foot Model with Medial and Lateral Midfoot and Forefoot Segments during Walking Gait. *J. Appl. Biomech.* **2016**, *32*, 608–613. [CrossRef] [PubMed]
31. Cavanagh, P.R. (Ed.) *Biomechanics of Distance Running*; Human Kinetics: Champaign, IL, USA, 1990.
32. Heiderscheit, B.C.; Hamill, J.; van Emmerik, R.E.A. Variability of Stride Characteristics and Joint Coordination among Individuals with Unilateral Patellofemoral Pain. *J. Appl. Biomech.* **2002**, *18*, 110–121. [CrossRef]
33. Hamill, J.; Haddad, J.M.; McDermott, W.J. Issues in Quantifying Variability from a Dynamical Systems Perspective. *J. Appl. Biomech.* **2000**, *16*, 407–418. [CrossRef]
34. Okunuki, T.; Koshino, Y.; Yamanaka, M.; Tsutsumi, K.; Igarashi, M.; Samukawa, M.; Saitoh, H.; Tohyama, H. Forefoot and Hindfoot Kinematics in Subjects with Medial Tibial Stress Syndrome during Walking and Running. *J. Orthop. Res.* **2019**, *37*, 927–932. [CrossRef] [PubMed]
35. Liu, Y.; Fernandez, J. Randomized Controlled Trial of Gastrocnemius Muscle Analysis Using Surface Electromyography and Ultrasound in Different Striking Patterns of Young Women's Barefoot Running. *Phys. Act. Health* **2024**, *8*, 223–233. [CrossRef]
36. Seneli, R.M.; Beschorner, K.E.; O'Connor, K.M.; Keenan, K.G.; Earl-Boehm, J.E.; Cobb, S.C. Foot Joint Coupling Variability Differences between Habitual Rearfoot and Forefoot Runners Prior to and Following an Exhaustive Run. *J. Electromyogr. Kinesiol.* **2021**, *57*, 102514. [CrossRef]
37. Fukano, M.; Fukubayashi, T.; Banks, S.A. Sex Differences in Three Dimensional Talocrural and Subtalar Joint Kinematics during Stance Phase in Healthy Young Adults. *Hum. Mov. Sci.* **2018**, *61*, 117–125. [CrossRef]
38. Matsumoto, Y.; Ogihara, N.; Kosuge, S.; Hanawa, H.; Kokubun, T.; Kanemura, N. Sex Differences in the Kinematics and Kinetics of the Foot and Plantar Aponeurosis during Drop-Jump. *Sci. Rep.* **2023**, *13*, 12957. [CrossRef]

biomechanics

MDPI

Article

Exploring the Effect of Prolonged Ankle Plantar-Flexed Standing on Postural Control, Balance Confidence, Falls Efficacy, and Perceived Balance in Older Adults

Daniel Craig Low

Centre for Physical Activity in Health and Disease, Brunel University of London, London UB8 3PH, UK; daniel.low@brunel.ac.uk

Abstract: Background/Objectives: Postural control describes our ability to maintain an upright position. This study explored the impact of prolonged ankle plantar-flexed standing on postural control variability and strategy in an older adult population. The ability to perceive balance change was also assessed via subjective balance-related variables. **Methods:** Twenty-four community-dwelling older adults were recruited via convenience sampling. Each participant completed a balance confidence and falls efficacy questionnaire at baseline. Five barefoot quiet standing trials on a force plate then followed (Timepoint 1). After this, the participants stood with their ankles in a plantar-flexed position for up to 7.5 min before completing another quiet standing trial on the force plate. Four further ankle plantar-flexed standing trials of 2 min were then completed, interspersed with quiet standing trials on a force plate (Timepoint 2). The balance confidence and falls efficacy questionnaires were then completed again. For measures of postural control variability (sway path length, root mean square [RMS], sway area) and strategy (fractal dimension), mean values for the five trials were calculated for Timepoints 1 and 2 separately. **Results:** The sway path length and RMS measures were significantly increased ($p < 0.05$) at Timepoint 2. However, the fractal dimension did not change. There was also no change in balance confidence or falls efficacy. **Conclusions:** The findings suggest that prolonged standing can impact measures of postural variability without a change in postural control strategy. Postural control change also occurred without a change in subjective balance measures, suggesting that the altered balance may not be practically significant or perceptible to the individual.

Keywords: postural variability; postural strategy; fatigue; older people

1. Introduction

Human postural control has been modelled as an inverted pendulum with muscular forces acting about the ankle joints to control the centre-of-mass movement [1,2]. Older adults who engage in prolonged activity involving the ankle plantar flexors may experience muscular fatigue, leading to increased postural sway area and sway path [3]; this is indicative of an increased fall risk [4,5].

The sway area and path length variables are linear measures of postural control that offer insight into postural variability [6]. A complimentary non-linear analysis, such as the calculation of fractal dimension, provides additional insight into postural control strategies [7,8]. However, previous studies of prolonged standing with older adult populations have not explored changes in control strategy. These studies, therefore, miss important insight into control mechanisms. Similarly, they tend to study the effect of volitional fatigue, which is less likely to be experienced by older adults. Furthermore, postural control data are inherently instable [9], which can impact the data's absolute and relative reliability [10]. Studies fail to offer interpretations of their findings in relation to the data's reliability, especially the Minimum Detectable Change (MDC), considered the smallest real difference that reflects true change rather than measurement error [11]. Consequently, these factors can impact the previous conclusions drawn.

The practical significance of postural control change is also missing in previous research. It is, therefore, unclear whether prolonged plantar-flexed standing leads to a perceptible change in balance, balance confidence, or falls efficacy. This is important given that a change in these subjective states relates to fear of falling, physical activity avoidance [12,13], and quality of life [14]. In fact, it is not clear whether there is a relationship between a change in these measures and a change in postural control variables.

This study sought to understand the effect of prolonged activity on the postural control variability and strategy of older adults. Furthermore, it aimed to evaluate whether there was any subjective balance change experienced and whether there were any relationships between changes in postural control and subjective balance measures. It was hypothesised that postural control will decrease following prolonged standing. It was also thought that balance confidence and falls efficacy will decrease following prolonged standing. Finally, it was hypothesised that a change in postural control measures will be positively associated with a perceived change in balance and falls efficacy and negatively associated with balance confidence.

2. Materials and Methods

A convenience sample of 24 healthy, community-living older adults (73.7 ± 6.8 years; Male/Female = 9/15) meeting the inclusion and exclusion criteria was recruited from the West London area of the United Kingdom. The inclusion criteria required participants (1) to be at least 60 years old and (2) to independently stand without an assistance device (e.g., cane and walkers). Older adult residents with cognitive impairment, deteriorated musculoskeletal or neurological function, and any medical disease history that impaired walking and balance (e.g., arthritis, diabetes, visuospatial deficits) were excluded. This study was approved by the Brunel University of London ethics committee (42477-A-Feb/2024-49776-1) in accordance with the Declaration of Helsinki. The number of participants was calculated based on the need to reach a statistical power of 0.8 with $p = 0.05$ and $d = 0.75$ [15]. A Cohen's d of 0.75 was chosen since it is considered large in gerontology [16], and these changes are described as 'grossly perceptible' [15] and, thus, could be considered important from a practical perspective.

At baseline, the participants undertook five barefoot quiet standing trials (20 s each) upon a force plate (Kistler, Winterthur, Switzerland; 100 Hz), standing with their feet shoulder width apart and arms by their sides, with visual fixation on a cross marked approximately 1.8 m in distance and 1.8 m in height on a wall. The participants then stood with their ankles plantar-flexed so that their heels were off the floor, for up to 7.5 min. After the completion of this time period or when they could not continue any longer, the participants stood on the force plate again for another 20 s quiet standing trial in the same stance. Plantar-flexed standing was then repeated four further times, each lasting two minutes, with a quiet standing trial collected after each. Each plantar-flexed stand was conducted next to the force plate to minimise the duration taken between the end of the task and the collection of postural control data. This, along with the quiet standing trial duration and conducting the plantar-flexed standing before each quiet standing trial, followed recommendations made to ensure fatigue remained present [17]. A chair was also placed in front of the participant so that the participants could use this to remain plantar-flexed standing for as long as possible.

Postural control variability was measured using the anterior–posterior (A-P), medio-lateral (M-L), and total sway path lengths; the A-P and M-L RMS and the RMS radius. and sway area (95% confidence area) were also calculated. These measures are all associated with the risk of falls and calculated using the position of the centre of pressure (COP) along with published formulae [18]. The fractal dimension was also calculated, using the 95% confidence area and total sway path length (Equation (1)), where N is number of sample points. A value of 1 indicated a completely stationary postural control signal, and a value of 2 indicated completely random postural control data [19]. The data were not filtered since this can remove natural variability, resulting in a loss of complexity [19]. The COP data were calculated within BioWare software (version 5.3, Kistler, Winterthur, Switzerland), and then all the postural control calculations were performed within Microsoft Excel (Microsoft Corporation, Redmond, WA, USA).

$$\frac{\log \mathrm{N}}{\log \mathrm{N} + \log\sqrt{\frac{4}{\pi} \times 95\%\ \text{confidence area}} - \log \text{total sway path length}} \quad (1)$$

Before baseline quiet standing trials and following the 5th plantar-flexion quiet standing trial, the participants completed the English Falls Efficacy Scale (FES-I) [13], which consists of 16 questions about the participants' concern of falling. The FES-I has excellent test–retest reliability and good internal consistency [13]; it also possesses good sensitivity in community-dwelling older adults [20], including for changes in physical function [21]. The sum of the ordinal data is calculated and interpreted as follows: those with scores of 16–19 have a low concern of falling; 20–27 indicates a moderate concern of falling; and above 28 indicates a high concern of falling [22]. The participants' balance confidence was also assessed using the Activity-Specific Balance Confidence short-version questionnaire (ABC) [12]. This consists of 6 questions with outcome scores presented as ordinal data. The scale possesses excellent test–retest reliability [23] and also has high internal consistency [24]. The ABC is related to physical function in community-dwelling older adults. The average score for the questions was determined and interpreted as follows: a score lower than 50 was indicative of low functioning, 50–80 was moderate functioning, and above 80 was high functioning [12].

Along with the FES-I and ABC, the participants also rated their perceived change in balance on a 15-point Generalised Rating of Change (GRC) question [25]. The self-report GRC is a single-item, recall-based questionnaire of global well-being and pain, based on change since an initial treatment encounter. The participants scored their global rating of change in balance compared with their baseline standing on a 15-point self-report Likert scale (from -7 to 7). A score of 1 to 7 suggests improvement; 0 suggests no change; and -1 to -7 indicates deterioration. The larger the value, the greater the degree of change. The outcome scores are ordinal and considered to have high face reliability [26]; they are also often used as an external standard of change in functional status [27]. The order for completing these questionnaires was varied across participants.

The effect of the prolonged standing was explored for each postural control variable by averaging the five baseline trials and comparing these with the average of the five trials collected post plantar-flexed standing; the averaging of multiple trials substantially improves the reliability of the data [28]. These were compared using paired sample t-tests. Similarly, Wilcoxon-signed rank tests were used to explore the differences in ABC and FES-I. The association between the change in postural control data and the change in balance confidence and falls efficacy (gain score) was determined using a Spearman's Rank Correlation Coefficient; these analyses were performed within SPSS software 29.0 (Version 29, IBM

Corp., Armonk, NY, USA). The size of the relationships was identified as weak when r = 0 to 0.3 or 0 to −0.3 (positive and negative relationship, respectively), moderate when r = 0.3 to 0.7 and −0.3 to −0.7 (positive and negative relationship, respectively), and strong when r = 0.7 to 1 and −0.7 to −1 (positive and negative relationship, respectively) [29]. Missing data were omitted from the calculations and reflected in the overall count of responses. Statistical significance was accepted at $p < 0.05$.

The relative reliability of the postural control data was assessed using the 5 baseline trials by calculating the average measures two-way random absolute agreement Intraclass Correlation Coefficient ($ICC_{2,5}$). These ICCs were interpreted using the following criteria: ICC of 0 to 0.5 indicates poor reliability; 0.5 to 0.75 is considered moderate reliability; 0.75 to 0.9 is considered good reliability; and above 0.9 is considered excellent reliability [30]. The standard error of the measurement (SEM) was determined using Equation (2), where SD was determined using Equation (3); the sum of squares total (SS_{total}) was provided within the ANOVA table provided by SPSS along with the ICC data [31].

$$\mathrm{SEM} = \mathrm{SD}\sqrt{1 - \mathrm{ICC}_{2,5}} \tag{2}$$

$$\sqrt{\mathrm{SS}_{\mathrm{total}} / (\mathrm{n} - 1)} \tag{3}$$

To assess whether any individual change in postural control data was real or due to chance, the MDC_{95} was determined using Equation (4).

$$\mathrm{MDC}_{95} = \mathrm{SEM} \times 1.96\sqrt{2} \tag{4}$$

3. Results

Based on the ABC, 15 participants were considered high functioning; 2 were low functioning; and 7 possessed a moderate level of functioning. Furthermore, 14 participants had a low concern about falling; 9 possessed moderate concern, and 1 participant had high concern.

All the participants performed five plantar-flexed standing trials, although seven participants could not sustain standing for the full 7.5 min duration for trial 1. However, all completed trials two to four, each lasting two minutes.

Prolonged standing resulted in a significantly greater A-P ($t(23) = -3.39$, $p = 0.003$, $d = 0.54$) and total sway path length ($t\ (23) = -3.35$, $p = 0.003$, $d = 0.39$); however, the sway length in the M-L direction did not change ($t(23) = -1.53$, $p = 0.140$, $d = 0.12$) (Figure 1)

The RMS in the A-P direction also significantly increased following plantar-flexed standing ($t(23) = -2.057$, $p = 0.051$), as did the RMS radius ($t(23) = 2.034$, $p = 0.054$). Conversely, the M-L RMS ($t(23) = -0.68$, $p = 0.50$) and sway area ($t(23) = -1.553$, $p = 0.134$) did not change, neither did the fractal dimension ($t(23) = -3.13$, $p = 0.757$) (Table 1).

Table 1. Mean, standard deviation, and Cohen's *d* for postural control measures taken at baseline (pre) and after plantar-flexed standing (post).

	Pre-Mean (SD)	Post-Mean (SD)	Cohen's *d*
RMS A-P (m) *	0.049 (±0.0009)	0.058 (±0.0021)	5.6
RMS M-L (m)	0.059 (±0.0009)	0.035 (±0.0012)	0.9
RMS radius (m) *	0.009 (±0.0124)	0.011 (±0.0141)	0.2
95% ellipse area (m^2)	0.0003 (±0.0001)	0.0004 (±0.0002)	0.6
Fractal dimension	1.72 (0.08)	1.73 (0.09)	0.1

* Significance < 0.05; A-P = anterior–posterior; M-L = medio-lateral; RMS = root mean square.

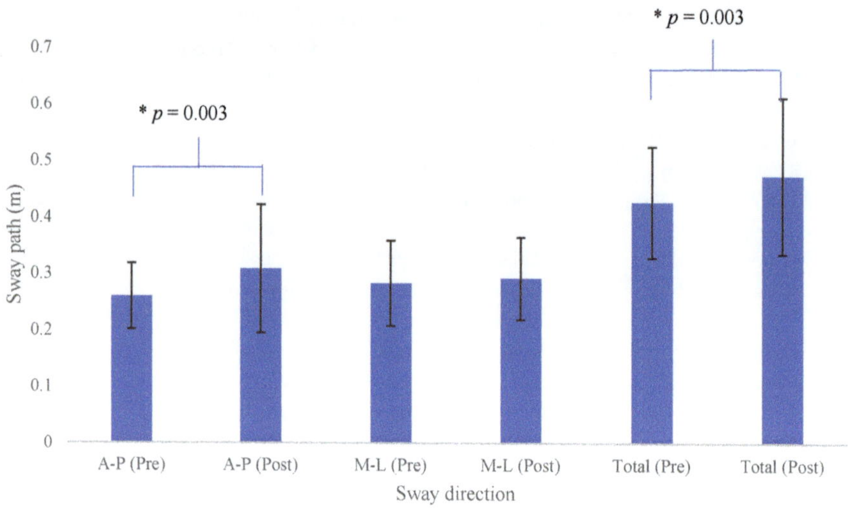

Figure 1. Comparison of total, medio-lateral (M-L), and anterior–posterior (A-P) sway path at baseline (pre) and after planar-flexed standing (post).

There was no difference in ABC when comparing baseline (median = 87.92, IQR = 22.92) with after plantar-flexed standing (median = 88.33, IQR = 26.04), ($z = -1.253$, $p = 0.210$); this was also true for FES-I (baseline median = 19, IQR = 3, after plantar-flexion standing median = 19, IQR = 4, $z = -1.190$, $p = 0.234$).

The GRC was 0 for 12 individuals, suggesting no perceived general change in balance. Seven individuals experienced a negative change in perceived balance (e.g., poorer balance), and five rated an improved balance (at least 1-point movement on the scale in either direction). Table 2 shows that the GRC was moderately and positively correlated with the change in A-P and total sway path length and RMS A-P. There were no other significant correlations for GRC nor were there any between sway data and ABC or falls efficacy gain scores.

Table 2. Correlation between postural control measurements and General Rating of Change in balance (GRC), Activity-Specific Balance Confidence (ABC), and Falls Efficacy Scale (FES-I).

		A-P Sway Length	M-L Sway Length	Total Sway Length	RMS A-P	RMS M-L	RMS Radius	Sway Area	Fractal Dimension
GRC	*R*	0.59	0.25	0.61	0.42	0.28	0.35	0.32	−0.13
	P	<0.01 *	0.25	<0.01 *	0.04 *	0.19	0.10	0.12	0.55
FES-I	*R*	0.02	0.18	0.07	−0.07	−0.08	−0.15	−0.02	0.05
	P	0.92	0.41	0.77	0.76	0.73	0.50	0.93	0.82
ABC	*R*	−0.31	−0.15	−0.32	0.14	−0.06	0.19	0.002	−0.16
	P	0.15	0.48	0.13	0.51	0.79	0.36	0.99	0.47

* Significance < 0.05; A-P = anterior–posterior; M-L = medio-lateral; RMS = root mean square.

The reliability ($ICC_{2,5}$) of the postural control measurements ranged between moderate ($ICC_{2,5} = 0.68$) to excellent ($ICC_{2,5} = 0.98$), except for A-P RMS which was poor and non-significant ($ICC_{2,5} = 0.27$). The MDC_{95} was represented as a percentage of the baseline average and ranged from 5.9% to 100% (Table 3).

Table 3. Relative and absolute reliability for postural control measurements.

	$ICC_{2,5}$	SS_{Total}	SD	SEM	MDC_{95}	$\%MDC_{95}$	Participants Exceeding MDC_{95} (n)
A-P sway path length	0.92 *	0.516	0.07	0.2	0.05	19.3%	9
M-L sway path length	0.98 *	0.685	0.08	0.1	0.03	10.6%	2
Total sway path length	0.96 *	1.265	0.10	0.2	0.06	14.0%	9
A-P RMS	0.27	0.0004	0.002	0.002	0.004	8.2%	9
M-L RMS	0.68 *	0.0002	0.001	0.0008	0.002	5.9%	10
RMS radius	0.95 *	0.21	0.013	0.003	0.008	88.9%	10
Sway area	0.69 *	0.000005	0.0002	0.0001	0.0003	100%	0
Fractal dimension	0.80 *	1.29	0.10	0.05	0.13	7.6%	0

* Significance < 0.01; ICC = Intraclass Correlation Coefficient; SS_{Total} = Sum of $squares_{Total}$; SD = standard deviation; SEM = standard error of measurement; MDC_{95} = Minimal Detectable $Change_{95}$; $\%MDC_{95}$ = Minimal Detectable $Change_{95}$ as a percentage of baseline mean average; A-P = anterior–posterior; M-L = medio-lateral; RMS = root mean square.

4. Discussion

This study explored the effect of prolonged activity on postural control. In partial agreement with the study hypothesis, prolonged activity decreased older adults' postural control for some measures, supporting previous observations [3,32]. The greater total sway path length and RMS radius along with greater movement in the A-P direction suggests decreased postural steadiness [33] and a greater risk of falling [18].

A decreased function of the plantar flexor muscles due to fatigue may underpin the observed changes in postural stability. These muscles control movement in the sagittal plane, across which stability changed. Conversely, in the M-L direction, the sway path length did not change, which is understandable given that the protocol did not target muscles that control movement in this direction. During prolonged standing, individuals can experience reduced blood flow [34], contributing to oxygen and nutrient depletion and the accumulation of lactic acid in the muscles [35]. Furthermore, during eccentric contraction, muscle fibres can experience high mechanical stress [36]; all of this can lead to muscle damage [35] and fatigue [37,38]. As a consequence, the muscles have a reduced mechanical power and force output [39], leading to a diminished ability to ensure postural stability. This peripheral fatigue can also impair motor control through weakened sensory integration and proprioception [40], important for effective postural control [41]. Whilst fatigue was not directly measured, the postural control changes observed are consistent with other studies in which individuals were fatigued [3,17,32]. However, it is also important to acknowledge that the prolonged standing may have also led to increased sway in response to pain or discomfort. Similarly, the fear of falling may have offered a psychological explanation for the change in sway [42], yet there were no differences in the subjective measures collected to suggest this.

The increase in sway paths and RMS was not accompanied by an increased sway area, an observation that is in contrast to those of Boyas et al. [3]. This may suggest that the ankle-stabilising muscles were able to retain sufficient joint stiffness to maintain the area in which the increased sway occurs. The difference in findings is likely due to greater fatigue induced by Boyas and colleagues, who used isokinetic contractions until failure. Isokinetic contractions can lead to fatigue more quickly than the isometric contractions used in the current study [43]. Sustaining activity until complete exhaustion is unlikely to occur in an older population, and, thus, this study provides evidence of a change in postural control under fatigue conditions more likely experienced in older adult populations.

Despite changes in postural variability, the signal complexity as indicated via the fractal dimension, was unchanged, suggesting that there was no reorganisation of afferents and no use of a new postural control strategy [19,44]; this is consistent with other studies in which increased postural variability has been observed [40,45]. Consequently, any change in proprioception due to the prolonged standing was insufficient for sensory reweighting to be required. Similarly, there was no need for increased activity from other ankle muscles or for a switch to a hip or multi-joint postural control strategy [1], which can occur when perturbations increase [46].

Whilst changes in the postural stability were shown, the prolonged activity also did not significantly change balance confidence or falls efficacy, measures considered important since they impact daily physical activity participation, fear of falling [12,13], and quality of life [14]. Practically, this suggests that postural control change did not impact the overall perception of balance, which may be because the postural control change was insufficient in magnitude. Consequently, the practical significance of the change in postural control is questionable, which is commonly not demonstrated in similar studies.

It is also important to consider the individual response to the task, which may somewhat account for the inconsistency between group postural control and perception changes. For various measures (sway area, A-P sway length, total sway length, and RMS A-P), there were significant moderate correlations with the GRC, which suggests that postural control change is responsive to changes in this subjective measure [47]. Therefore, for some individuals, the size of change in postural control was notable, whereas, for others, it was not. However, these postural control changes were not related to a change in confidence or efficacy, suggesting that other factors may be more important when evaluating the perception of fall risk or movement confidence.

Group differences may have also been influenced by variability within the data, affecting their reliability. The relative reliability of the sway path data was excellent, whilst that of sway area was good and larger than that presented previously by some (ICC = 0.22, [10]) but by not others using similar populations (ICC = 0.92, [48]). The RMS radius also had excellent reliability, which was larger than that previously presented (ICC = 0.82, [48]). The RMS in the ML direction had moderate reliability; this was higher than the reliability presented by Lafond et al. [10] and similar to that by Swanenburg et al. [49]. In general, this suggests that there was confidence in these data measured. Conversely, the RMS-AP data showed poor reliability, which is in line with other studies [10,49] but which may impact the ability to draw faithful conclusions from these data.

Despite generally good-to-excellent reliability, fewer than half of the individuals experienced a change that was beyond the MDC_{95}. Therefore, the group difference may have resulted from a number of individual changes that were real, combined with others that were the result of chance. Consequently, this may explain why a corresponding change in the ABC and FES-I data was not observed. The reasons for inter-individual differences may include differences in frailty or sensory system functionality that were either insufficient for clinical diagnosis or were undiagnosed or unreported by the participants. Differences in body mass index are shown to impact postural control [50], which, along with differences in muscle fatigue tolerance, may also be influential. Differences in the baseline ABC and FES-I may also suggest that individual function and confidence may influence the postural control and perceptions reported. These factors are rarely considered in the study of postural control; thus, future studies should be more considerate of these factors when studying change.

Collectively, these results indicate the importance of measuring perceptual data alongside reliability data when interpreting postural control in older adults. Furthermore, for measures deemed responsive, future research should determine the Minimal Important Change (MIC) for each [47], which will allow the assessment of the meaningfulness of individual changes. It is important to note that this is not the same as the MDC, which is the minimum change required to be real; instead, it is the size of change considered important for the individual.

A study limitation is that the relative fatigue of each participant was not measured. This can be determined using dynamometry or electrical stimulation, yet this was not practically possible given the recommendations to ensure fatigue is present [17]. The RPE could have also been measured although the maximum RPE at fatigue is often not 20 (i.e., maximum score) and can be dependent on the individual and the exercise; recording the RPE to make comparisons of individuals within and between studies is, therefore, problematic, and there is also generally no correlation between fatigue-induced RPE and change in balance [51]. The angle of the plantar-flexion stance was also not monitored, which may impact the fatigue generated; thus, motion analysis could be used to monitor this in future research. Electromyography could also be used to explore a change in muscle activation that may be indicative of peripheral muscle fatigue [52]. A larger sample size and greater task intensity may have also resulted in differences and correlations being shown for those variables observed as insignificant.

5. Conclusions

In conclusion, older adults undertaking prolonged activity can experience increased postural variability without a change in sway area or control strategy. This change in variability also occurred without a group change in balance confidence or falls efficacy. The change in postural control variability was related to a perceived change in balance but not in balance confidence or falls efficacy. This study also shows the importance of understanding individual change when interpreting postural control data in group comparisons, which is often missing in similar research.

Funding: This research received no external funding.

Institutional Review Board Statement: This study was conducted in accordance with the Declaration of Helsinki and approved by the Institutional Review Board (or Ethics Committee) of Brunel University London (protocol code 42477 and 2nd February 2024).

Informed Consent Statement: Informed consent was obtained from all the subjects involved in this study.

Data Availability Statement: The datasets presented in this article are not readily available because this was not a feature of the ethical approval.

Conflicts of Interest: The author declares no conflicts of interest.

Abbreviations

The following abbreviations are used in this manuscript:

A-P	Anterior–posterior
M-L	Medio-lateral
RMS	Root mean square
GRC	Generalised Rating of Change
ICC	Intraclass Correlation Coefficient
SS_{total}	Sum of squares total
MDC	Minimal Detectable Change

References

1. Winter, D.A. Human balance and posture control during standing and walking. *Gait Posture* **1995**, 3, 193–214. [CrossRef]
2. Winter, D.A.; Patla, A.E.; Rietdyk, S.; Ishac, M.G. Ankle muscle stiffness in the control of balance during quiet standing. *J. Neurophysiol.* **2001**, 85, 2630–2633. [CrossRef] [PubMed]

3. Boyas, S.; Medd, E.R.; Beaulieu, S.; Boileau, A.; Lajoie, Y.; Bilodeau, M. Older and young adults adopt different postural strategies during quiet bipedal stance after ankle plantarflexor fatigue. *Neurosci. Lett.* **2019**, *701*, 208–212. [CrossRef] [PubMed]
4. Howcroft, J.; Lemaire, E.D.; Kofman, J.; McIlroy, W.E. Elderly fall risk prediction using static posturography. *PLoS ONE* **2017**, *12*, e0172398. [CrossRef]
5. Merlo, A.; Zemp, D.; Zanda, E.; Rocchi, S.; Meroni, F.; Tettamanti, M.; Recchia, A.; Lucca, U.; Quadri, P. Postural stability and history of falls in cognitively able older adults: The Canton Ticino study. *Gait Posture* **2012**, *36*, 662–666. [CrossRef]
6. Smith, B.A.; Stergiou, N.; Ulrich, B.D. Lyapunov exponent and surrogation analysis of patterns of variability: Profiles in new walkers with and without down syndrome. *Motor Control* **2010**, *14*, 126–142. [CrossRef]
7. Kiefer, A.W.; Armitano-Lago, C.N.; Cone, B.L.; Bonnette, S.; Rhea, C.K.; Cummins-Sebree, S.; Riley, M.A. Postural control development from late childhood through young adulthood. *Gait Posture* **2021**, *86*, 169–173. [CrossRef]
8. Doyle, T.L.; Newton, R.U.; Burnett, A.F. Reliability of traditional and fractal dimension measures of quiet stance center of pressure in young, healthy people. *Arch. Phys. Med. Rehabil.* **2005**, *86*, 2034–2040. [CrossRef]
9. Rhea, C.K.; Silver, T.A.; Hong, S.L.; Ryu, J.H.; Studenka, B.E.; Hughes, C.M.; Haddad, J.M. Noise and complexity in human postural control: Interpreting the different estimations of entropy. *PLoS ONE* **2011**, *6*, e17696. [CrossRef]
10. Lafond, D.; Corriveau, H.; Herbert, R.; Prince, F. Intrasession reliability of center of pressure measures of postural steadiness in healthy elderly people. *Arch. Phys. Med. Rehabil.* **2004**, *85*, 896–901. [CrossRef]
11. Salehi, R.; Ebrahimi, T.I.; Esteki, A.; Maroufi, N.; Parnianpour, M. Test-retest reliability and minimal detectable change for center of pressure measures of postural stability in elderly subjects. *Med. J. Islam. Repub. Iran* **2010**, *23*, 224–232.
12. Myers, A.M.; Fletcher, P.C.; Myers, A.H.; Sherk, W. Discriminative and evaluative properties of the activities-specific balance confidence (ABC) scale. *J. Gerontol. A Biol. Sci. Med. Sci.* **1998**, *53*, M287–M294. [CrossRef] [PubMed]
13. Yardley, L.; Beyer, N.; Hauer, K.; Kempen, G.; Piot-Ziegler, C.; Todd, C. Development and initial validation of the Falls Efficacy Scale International (FES-I). *Age Ageing* **2005**, *34*, 614–619. [CrossRef] [PubMed]
14. Legters, K.; Verbus, N.B.; Kitchen, S.; Tomecsko, J.; Urban, N. Fear of falling, balance confidence and health-related quality of life in individuals with postpolio syndrome. *Physiother. Theory Pract.* **2006**, *22*, 127–135. [CrossRef]
15. Cohen, J. *Statistical Power Analysis for the Behavioural Sciences*, 2nd ed.; Erlbaum: Hillsdale, NJ, USA, 1988.
16. Brydges, C.R. Effect size guidelines, sample size calculations, and statistical power in gerontology. *Innov. Aging* **2019**, *3*, igz036. [CrossRef]
17. Vuillerme, N.; Forestier, N.; Nougier, V. Attentional demands and postural sway: The effect of the calf muscles fatigue. *Med. Sci. Sports Exerc.* **2002**, *34*, 1907–1912. [CrossRef]
18. Quijoux, F.; Nicolaï, A.; Chairi, I.; Bargiotas, I.; Ricard, D.; Yelnik, A.; Oudre, L.; Bertin-Hugault, F.; Vidal, P.P.; Vayatis, N.; et al. A review of center of pressure (COP) variables to quantify standing balance in elderly people: Algorithms and open-access code. *Physiol. Rep.* **2021**, *9*, e15067. [CrossRef]
19. Doyle, T.L.; Dugan, E.L.; Humphries, B.; Newton, R.U. Discriminating between elderly and young using a fractal dimension analysis of centre of pressure. *Int. J. Med. Sci.* **2004**, *1*, 11–20. [CrossRef]
20. Hauer, K.A.; Kempen, G.I.; Schwenk, M.; Yardley, L.; Beyer, N.; Todd, C.; Oster, P.; Zijlstra, G.A. Validity and sensitivity to change of the Falls Efficacy Scales International to assess fear of falling in older adults with and without cognitive impairment. *Gerontology* **2011**, *57*, 462–472. [CrossRef]
21. McColl, L.; McMeekin, P.; Poole, M.; Parry, S.W. Is fear of falling key to identifying gait and balance abnormalities in community-dwelling older adults? Protocol of a mixed-methods approach. *BMJ Open* **2022**, *12*, e067040. [CrossRef]
22. Delbaere, K.; Close, J.C.; Mikolaizak, A.S.; Sachdev, P.S.; Brodaty, H.; Lord, S.R. The Falls Efficacy Scale International (FES-I). A comprehensive longitudinal validation study. *Age Ageing* **2010**, *39*, 210–216. [CrossRef] [PubMed]
23. Schepens, S.; Goldberg, A.; Wallace, M. The short version of the Activities-specific Balance Confidence (ABC) scale: Its validity, reliability, and relationship to balance impairment and falls in older adults. *Arch. Gerontol. Geriatr.* **2010**, *51*, 9–12. [CrossRef] [PubMed]
24. Peretz, C.; Herman, T.; Hausdorff, J.M.; Giladi, N. Assessing fear of falling: Can a short version of the Activities-specific Balance Confidence scale be useful? *Mov. Disord.* **2006**, *21*, 2101–2105. [CrossRef]
25. Wright, A.A.; Abbott, J.H.; Baxter, D.; Cook, C. The ability of a sustained within-session finding of pain reduction during traction to dictate improved outcomes from a manual therapy approach on patients with osteoarthritis of the hip. *J. Man. Manip. Ther.* **2010**, *18*, 166–172. [CrossRef] [PubMed]
26. Kamper, S.J.; Maher, C.G.; Mackay, G. Global rating of change scales: A review of strengths and weaknesses and considerations for design. *J. Man. Manip. Ther.* **2009**, *17*, 163–170. [CrossRef]
27. Low, D.C.; Walsh, G.S. The minimal important change for measures of balance and postural control in older adults: A systematic review. *Age Ageing* **2022**, *51*, afac284. [CrossRef]
28. Rafał, S.; Janusz, M.; Wiesław, O.; Robert, S. Test-retest reliability of measurements of the center of pressure displacement in quiet standing and during maximal voluntary body leaning among healthy elderly men. *J. Hum. Kinet.* **2011**, *28*, 15–23. [CrossRef]

29. Ratner, B. The correlation coefficient: Its values range between +1/−1, or do they? *J. Target. Meas. Anal. Mark.* **2009**, *17*, 139–142. [CrossRef]
30. Koo, T.K.; Li, M.Y. A guideline of selecting and reporting intraclass correlation coefficients for reliability research. *J. Chiropr. Med.* **2016**, *15*, 155–163. [CrossRef]
31. Weir, J.P. Quantifying test-retest reliability using the intraclass correlation coefficient and the SEM. *J. Strength. Cond. Res.* **2005**, *19*, 231–240.
32. Freitas, S.M.; Wieczorek, S.A.; Marchetti, P.H.; Duarte, M. Age-related changes in human postural control of prolonged standing. *Gait Posture* **2005**, *22*, 322–330. [CrossRef] [PubMed]
33. Sozzi, S.; Ghai, S.; Schieppati, M. Incongruity of geometric and spectral markers in the assessment of body sway. *Front. Neurol.* **2022**, *13*, 929132. [CrossRef] [PubMed]
34. Waters, T.R.; Dick, R.B. Evidence of health risks associated with prolonged standing at work and intervention effectiveness. *Rehabil. Nurs.* **2015**, *40*, 148–165. [CrossRef] [PubMed]
35. Allen, D.G.; Lamb, G.D.; Westerblad, H. Skeletal muscle fatigue: Cellular mechanisms. *Physiol. Rev.* **2008**, *88*, 287–332. [CrossRef]
36. Hody, S.; Croisier, J.L.; Bury, T.; Rogister, B.; Leprince, P. Eccentric Muscle Contractions: Risks and Benefits. *Front. Physiol.* **2019**, *10*, 536. [CrossRef]
37. Acaster, S.; Dickerhoof, R.; DeBusk, K.; Bernard, K.; Strauss, W.; Allen, L.F. Qualitative and quantitative validation of the FACIT-fatigue scale in iron deficiency anemia. *Health Qual. Life Outcomes* **2015**, *13*, 60. [CrossRef]
38. Hepple, R.T. The Role of O_2 Supply in Muscle Fatigue. *Can. J. Appl. Physiol.* **2002**, *27*, 56–69. [CrossRef]
39. Nocella, M.; Colombini, B.; Benelli, G.; Cecchi, G.; Bagni, M.A.; Bruton, J. Force decline during fatigue is due to both a decrease in the force per individual cross-bridge and the number of cross-bridges. *J. Physiol.* **2011**, *589*, 3371–3381. [CrossRef]
40. Penedo, T.; Polastri, P.F.; Rodrigues, S.T.; Santinelli, F.B.; Costa, E.C.; Imaizumi, L.F.I.; Barbieri, R.A.; Barbieri, F.A. Motor strategy during postural control is not muscle fatigue joint-dependent, but muscle fatigue increases postural asymmetry. *PLoS ONE* **2021**, *16*, e0247395. [CrossRef]
41. Lee, J.W.; Park, J.S. The Correlation between Proprioception and Postural Control in Healthy Adults. *Iran. J. Public Health* **2022**, *51*, 2360–2361. [CrossRef]
42. Adkin, A.; Frank, J.S.; Jog, M.S. Fear of falling and postural control in Parkinson's disease. *Mov. Disord.* **2003**, *18*, 496–502. [CrossRef] [PubMed]
43. Bisson, E.J.; Remaud, A.; Boyas, S.; Lajoie, Y.; Bilodeau, M. Effects of fatiguing isometric and isokinetic ankle exercises on postural control while standing on firm and compliant surfaces. *J. Neuroeng. Rehabil.* **2012**, *9*, 39. [CrossRef] [PubMed]
44. Doherty, C.; Bleakley, C.; Hertel, J.; Caulfield, B.; Ryan, J.; Delahunt, E. Postural control strategies during single limb stance following acute lateral ankle sprain. *Clin. Biomech.* **2014**, *29*, 643–649. [CrossRef] [PubMed]
45. Boyas, S.; Hajj, M.; Bilodeau, M. Influence of ankle plantarflexor fatigue on postural sway, lower limb articular angles, and postural strategies during unipedal quiet standing. *Gait Posture* **2013**, *37*, 547–551. [CrossRef]
46. Horak, F.B.; Shupert, C.L.; Mirka, A. Components of postural dyscontrol in the elderly: A review. *Neurobiol. Aging* **1989**, *10*, 727–738. [CrossRef]
47. Terwee, C.B.; Peipert, J.D.; Chapman, R.; Lai, J.S.; Terluin, B.; Cella, D.; Griffiths, P.; Mokkink, L.B. Minimal important change (MIC): A conceptual clarification and systematic review of MIC estimates of PROMIS measures. *Quality Life Res.* **2021**, *30*, 2729–2754. [CrossRef]
48. Lin, D.; Seol, H.; Nussbaum, M.A.; Madigan, M.L. Reliability of COP-based postural sway measures and age-related differences. *Gait Posture* **2008**, *28*, 337–342. [CrossRef]
49. Swanenburg, J.; de Bruin, E.D.; Favero, K.; Uebelhart, D.; Mulder, T. The reliability of postural balance measures in single and dual tasking in elderly fallers and non-fallers. *BMC Musculoskelet. Disord.* **2008**, *9*, 162. [CrossRef]
50. Ku, P.X.; Abu Osman, N.A.; Yusof, A.; Wan Abas, W.A. Biomechanical evaluation of the relationship between postural control and body mass index. *J. Biomech.* **2012**, *45*, 1638–1642. [CrossRef]
51. Jo, D.; Bilodeau, M. Rating of perceived exertion (RPE) in studies of fatigue-induced postural control alterations in healthy adults: Scoping review of quantitative evidence. *Gait Posture* **2021**, *90*, 167–178. [CrossRef]
52. Kallenberg, L.A.C.; Schulte, E.; Disselhorst-Klug, C.; Hermens, H.J. Myoelectric manifestations of fatigue at low contraction levels in subjects with and without chronic pain. *J. Electromyogr. Kinesiol.* **2007**, *17*, 264–274. [CrossRef]

biomechanics

MDPI

Systematic Review

Gait Biomechanical Parameters Related to Falls in the Elderly: A Systematic Review

Jullyanne Silva [1], Tiago Atalaia [2], João Abrantes [3] and Pedro Aleixo [1,*]

1 Centro de Investigação em Desporto, Educação Física, Exercício e Saúde (CIDEFES), Universidade Lusófona, 1749-024 Lisbon, Portugal; jullyanne_silva@hotmail.com
2 Centro de Investigação em Ciências e Tecnologias da Saúde (CrossI&D), Escola Superior de Saúde da Cruz Vermelha Portuguesa-Lisboa, Physiotherapy, 1300-125 Lisbon, Portugal; tatalaia@esscvp.eu
3 Centre for Research in Applied Communication, Culture, and New Technologies (CICANT), Universidade Lusófona, 1749-024 Lisbon, Portugal; joao.mcs.abrantes@ulusofona.pt
* Correspondence: pedro.aleixo@ulusofona.pt

Abstract: According to the World Health Organization, one-third of elderly people aged 65 or over fall annually, and this number increases after 70. Several gait biomechanical parameters were associated with a history of falls. This study aimed to conduct a systematic review to identify and describe the gait biomechanical parameters related to falls in the elderly. MEDLINE Complete, Cochrane, Web of Science, and CINAHL Complete were searched for articles on 22 November 2023, using the following search sentence: (gait) AND (fall*) AND ((elder*) OR (old*) OR (senior*)) AND ((kinematic*) OR (kinetic*) OR (biomechanic*) OR (electromyogram*) OR (emg) OR (motion analysis*) OR (plantar pressure)). This search identified 13,988 studies. From these, 96 were selected. Gait speed, stride/step length, and double support phase are gait biomechanical parameters that differentiate fallers from non-fallers. Fallers also tended to exhibit higher variability in gait biomechanical parameters, namely the minimum foot/toe clearance variability. Although the studies were scarce, differences between fallers and non-fallers were found regarding lower limb muscular activity and joint biomechanics. Due to the scarce literature and contradictory results among studies, it is complex to draw clear conclusions for parameters related to postural stability. Minimum foot/toe clearance, step width, and knee kinematics did not differentiate fallers from non-fallers.

Keywords: gait; falls; spatiotemporal parameters; kinematics; kinetics; systematic review

1. Introduction

According to the World Health Organization [1], the occurrence of falls in the elderly population is an important public health problem, representing the second leading cause of unintentional injury deaths worldwide. Several individual consequences of falls are described in the literature: reduced quality of life [2], psychological effects, such as increased fear of falling and loss of confidence [3], and fractures [4]. The elderly are the group with the highest incidence and worst consequences of falls, which increase with aging [5]. On the other hand, it is also important to highlight the economic burden of falls for families, communities, and society, e.g., falls among the elderly in the United States of America resulted in substantial medical costs [6].

Falls in the elderly are dependent on complex multifactorial risks such as (1) intrinsic risks, e.g., muscle weakness, stability disorders, functional and cognitive impairment, and visual deficits; (2) extrinsic risks, e.g., prescription of four or more medications; and (3) environmental risks, e.g., poor lighting or rugs that slide [7]. Focusing our attention on intrinsic factors, it is important to highlight that the subjects' functional capacity and motor control play an important role in falls. This is particularly important during walking, one of the activities of daily life in which falls are most prevalent [8]. Moreover, the increment of knowledge about gait biomechanics related to falls (i.e., objective data of functional capacity and motor control) may help in the identification of subjects that present a high risk of falls and also in the development of interventions to decrease this risk [9]. In this context, tripping and slipping are the most frequent causes of falls during gait [10]. Regarding tripping, the minimum toe clearance (MTC) emerged as one gait biomechanical parameter related to this kind of fall. A previous systematic review [11] addressed this issue and found no differences between fallers and non-fallers regarding MTC, although the literature found was scarce. Furthermore, this systematic review also found that fallers had greater variability in MTC. Naturally, the biomechanical parameters related to slipping are different. In this way, the heel anteroposterior (AP) velocity at heel strike has been a biomechanical gait parameter related to slips [12]. Finally, the gait spatiotemporal parameters were also associated with a history of falls, namely gait speed and cadence [13,14]; stride length, double support phase duration, and variability in the stride length and swing time [14]; and variability in the step length and double-support phase [13].

Systematic reviews can synthesize the state of knowledge on a specific topic, allowing the identification of knowledge gaps that can constitute priorities for future research [15]. In our study, the synthesis of knowledge on the gait biomechanical parameters related to falls in the elderly may be useful for the stratification of severity (i.e., risk of fall) in future research and also helpful in developing more tailored and suitable assessments and interventions for this population [16]. Nonetheless, to the best of our knowledge, no systematic review has approached this issue. Therefore, the aim of this systematic review was to identify and describe the gait biomechanical parameters related to falls in elderly populations.

2. Materials and Methods

This systematic review was developed in accordance with the Preferred Items for Reporting for Systematic Reviews and Meta-Analysis (PRISMA) statement [17]. This review was registered in PROSPERO (ID–CRD42021271511).

2.1. Eligibility Criteria

The present review included studies according to the following inclusion criteria: (1) articles comparing elderly fallers and non-fallers on data from gait biomechanical analyses (the distinction between fallers and non-fallers can occur through retrospective studies, based on a previous history of falls, and prospective studies based on a follow-up of a fall's occurrence); (2) articles that during their methodological set-up induced falls and compared the elderly who fell with the elderly who recovered, regarding gait biomechanical data. In the scope of this study, gait biomechanical analyses also comprised analyses on uneven terrain. Furthermore, data from biomechanical gait analyses included spatiotemporal, kinematic, kinetic, electromyography (EMG), and plantar pressure data. The WHO definition of elderly was also considered, i.e., subjects aged 60 or over [18]. The following exclusion criteria were also defined: (1) articles assessing subjects under 60 years (or the mean age of the subjects less one standard deviation lower than 60 years); (2) articles assessing subjects that use any walking aid, with neurological or osteoarticular disease (e.g., stroke, Parkinson's disease, polyneuropathy, osteoarthritis, or rheumatoid arthritis), with dementia, who are amputees, or who are blind; (3) articles assessing subjects during dual-task, ascending or descending stairs, turning, or obstacle-crossing; and (4) case reports, reviews, or dissertations. No restrictions were imposed regarding language or publication date.

2.2. Search Strategy and Selection Process

Systematic review was conducted independently by two researchers (J.S. and P.A.), using the following protocol: (1) a comprehensive search of articles was made on MEDLINE Complete, Cochrane, Web of Science, and CINAHL Complete for articles published until 22 November 2023, using the following search sentence- (gait) AND (fall*) AND ((elder*) OR (old*) OR (senior*)) AND ((kinematic*) OR (kinetic*) OR (biomechanic*) OR (electromyogram*) OR (emg) OR (motion analysis*) OR (plantar pressure)); (2) exclusion of duplicates—Mendeley was used to manage all references, removing duplicates; (3) selection of articles by title and abstract; (4) screening of articles by analyzing the full text; and (5) hand search for relevant articles. In the third and fourth points of the protocol, a third reviewer (T.A.) would be consulted if there were any disagreements between the two reviewers.

2.3. Data Extraction and Synthesis

Data were extracted by one reviewer (J.S.) using a predefined form: (1) authors and year of publication; (2) inclusion and exclusion criteria and definition of fall; (3) sample characteristics (sample size and sociodemographic data—age, gender, and type of population); (4) gait assessment; and (5) results (gait biomechanical parameters related or not related to falls). All information was cross-checked by a second reviewer (P.A.).

2.4. Risk of Bias Assessment

In this systematic review, the Quality Assessment Tool for Quantitative Studies, developed by the Effective Public Health Practice Project [19], was used to assess the methodological quality of studies. This assessment focused on 6 domains: (1) selection bias; (2) study design; (3) confounders; (4) blinding; (5) data collection method; and (6) withdrawals and dropouts. Each domain was evaluated with the following classification: "1" corresponds to a strong report in that domain; "2" corresponds to a reasonable report in that domain; and "3" corresponds to a weak report in that domain. The articles that met the inclusion criteria were assessed independently by two researchers (J.S. and P.A.). Any disagreements were resolved with a consensus discussion between them. If disagreements remained after discussion, a third reviewer was consulted (T.A.).

3. Results

A total of 13,988 records were found from the following databases: MEDLINE Complete and CINAHL Complete (7144), Cochrane (887), and Web of Science (5957). Additional articles were also found by manual searching (12). Removal of duplicates resulted in 9425 eligible articles. After this first selection, 8512 articles were excluded by determining that their titles and abstracts were not relevant or did not meet the inclusion criteria. In this way, the full text of 913 records was reviewed. The eligibility process resulted in the inclusion of 96 articles in this systematic review. A flow diagram summarizing this selection process is shown in Figure 1.

3.1. Characteristics of the Selected Studies

This review included 96 studies: 86 studies [14,20–104] comparing elderly fallers and non-fallers (Table 1) and 12 studies [103–114] that induced falls during their methodological set-up and compared the elderly who fell with the elderly who recovered (Table 2).

Table 1. Characteristics and data of the studies that compared fallers' and non-fallers' gait.

Study	Inclusion and/or Exclusion Criteria Definition of Fall	Sample Characteristics	Gait Assessment	Gait Parameters Related to Falls (Fallers vs. Non-Fallers)	Gait Parameters Not Related to Falls (Fallers vs. Non-Fallers)
Heitmann et al., 1989 [20]	Exclusion criteria: Parkinson's disease; multiple sclerosis; or residual effects from a stroke. Inclusion criteria: able to walk 90 feet without an assistive device and to be independent in activities of daily living. Definition of fall: not reported.	Community-dwelling elderly women: 26 fallers (≥1 fall in past year; 75.1 ± 7.7 years). 84 non-fallers (73.1 ± 7.0 years).	Subjects walked on paper walkways making ink prints for step-width measurements. 3 trials were performed; the best one was used for analysis.		Step width (cm): 7.44 vs. 6.54; step width variability (SD; cm): 3.60 vs. 3.35.
Gehlsen & Whaley, 1990 [21]	Exclusion criteria: uncontrolled hypertension; angina pectoris; recent myocardial infarction; or disabling injury to legs and back. Definition of fall: not reported.	Community-dwelling elderly: 25 fallers (≥1 fall in past 10 months; 7 males; 72.4 ± 4.7 years). 30 non-fallers (12 males; 71.3 ± 4.4 years).	Subjects walked on a treadmill at 4 km/h and 6 km/h. They were filmed by two cameras (64 Hz; sagittal and frontal planes). MTC was analyzed.	6 km/h: heel width (cm): 7.39 vs. 6.41.	4 km/h: heel width (cm): 7.77 vs. 7.19. 4 km/h and 6 km/h: stride length (m): 0.58 vs. 0.59; 0.73 vs. 0.72; MTC (cm): 1.15 vs. 0.70; 0.77 vs. 0.68; single support phase (s): 0.49 vs. 0.50; 0.45 vs. 0.44; double support phase (s): 0.14 vs. 0.15; 0.11 vs. 0.11; swing phase (s): 0.35 vs. 0.35; 0.34 vs. 0.34; cadence (stride/s): 2.05 vs. 2.01; 2.26 vs. 2.21; hip displacement (°): 65–104 vs. 64–105; knee displacement (°): 9–62 vs. 8–63; 7–63 vs. 8–62; ankle displacement (°): 98–120 vs. 97–120; 103–125 vs. 99–122.
Feltner et al., 1994 [22]	Exclusion and inclusion criteria: not reported. Definition of fall: event that results in a subject coming to rest inadvertently on the ground.	Community-dwelling elderly women: 6 fallers (≥1 fall in past year; 71.7 ± 2.6 years). 11 non-fallers (74.4 ± 1.7 years).	Subjects walked at their preferred gait speed across an 8.2 m walkway filmed by two cameras (60 Hz; sagittal and frontal planes). At least 3 trials were collected, and the trial with a complete stride in the side view camera footage was used for analysis. AP and ML velocities of CoM were calculated.		Stride length (m): 1.12 vs. 1.16; right and left step length (m): 0.57 vs. 0.60, 0.54 vs. 0.56; step width (m): 0.22 vs. 0.22; stride time (s): 1.05 vs. 1.02; right and left step time (s): 0.59 vs. 0.53, 0.46 vs. 0.50; single support (%): 69.8 vs. 68.5; swing phase (%): 31.1 vs. 32.9; CoM AP velocity (m/s): 1.08 vs. 1.14; CoM ML velocity (m/s): −0.19 vs. −0.20; minimum and maximum hip position (°): −6 vs. −10; 26 vs. 26; minimum and maximum knee position (°): 175 vs. 175; 118 vs. 116; minimum and maximum ankle position (°): 12 vs. 8; −12 vs. −14; width of the base of support (m): 0.14 vs. 0.15.

Table 1. *Cont.*

Study	Inclusion and/or Exclusion Criteria Definition of Fall	Sample Characteristics	Gait Assessment	Gait Parameters Related to Falls (Fallers vs. Non-Fallers)	Gait Parameters Not Related to Falls (Fallers vs. Non-Fallers)
Wolfson et al., 1995 [23]	Exclusion criteria: terminal illness; severe dementia; non-ambulatory status; required use of a walker; amputations; severe arthritis; major impairment due to neurologic disease; or episodes of loss of consciousness. Definition of fall: not reported.	Community-dwelling elderly: 18 fallers (≥2 falls in past year; mean age 82.2 years). 15 non-fallers (mean age 84.6 years).	Not reported.	Gait speed (m/s): 0.37 vs. 0.64; stride length (m): 0.53 vs. 0.82.	
Maki, 1997 [24]	Inclusion criteria: able to walk 10 m with or without a walking aid and understand verbal instructions. Definition of fall: event that results in a subject coming to rest inadvertently on the ground.	Community-dwelling elderly: 43 fallers (≥1 fall in 1-year follow-up; 8 males; 82.8 ± 6.2 years). 32 non-fallers (6 males; 81.0 ± 6.7 years).	Subjects walked with their own footwear at their preferred gait speed across an 8 m walkway. Four trials were filmed but only the last two were included in the analysis.	Stride length variability (SD): higher values in fallers; double support phase variability (SD): higher values in fallers; gait speed variability (SD): higher values in fallers.	Stride length; stride time; double support phase; gait speed; stride width variability (SD); stride time variability (SD).
Lee & Kerrigan, 1999 [25]	Exclusion criteria: musculoskeletal, neurological, or cardiac diseases. Definition of fall: event that results in a subject coming to rest inadvertently on the ground or other lower level.	Community-dwelling elderly: 15 fallers (≥2 falls in past 6 months; 7 males; 77.0 ± 9.0 years). 15 non-fallers (8 males; 75.0 ± 5.0 years).	Subjects walked barefoot or with their shoes at a preferred gait speed across a 30-foot walkway. Kinematic data (using a 4-camera optoelectronic motion analysis system at 100 Hz) and ground reaction forces (using 2 force plates) were collected during 3 trials.	Gait speed (m/s): 0.41 vs. 0.82; cadence (steps/s): 86 vs. 111; step length (m): 0.24 vs. 0.40; hip flexion moment (Nm/kg): 0.96 vs. 0.44; hip adduction moment (Nm/kg): 1.49 vs. 0.69; knee varus moment (Nm/kg): 0.86 vs. 0.33; knee extension moment (Nm/kg): 0.40 vs. 0.21; ankle dorsiflexion moment (Nm/kg): 1.59 vs. 0.91; ankle plantarflexion moment (Nm/kg): 0.075 vs. 0.139; ankle eversion moment (Nm/kg): 0.43 vs. 0.13; knee power absorption (W/kg): 0.98 vs. 1.66; ankle power absorption (W/kg): 0.76 vs. 0.41.	Hip extension moment (Nm/kg): 0.67 vs. 0.74; knee flexion moment (Nm/kg): 0.59 vs. 0.46; ankle inversion moment (Nm/kg): 014 vs. 0.07; hip power generation (W/kg): 1.23 vs. 1.18; knee power generation (W/kg): 0.54 vs. 0.55; ankle power generation (W/kg): 2.03 vs. 2.04; hip power absorption (W/kg): 0.44 vs. 0.61.
Nelson et al., 1999 [26]	Inclusion criteria: independent subjects. Definition of fall: not reported.	Community-dwelling elderly: 11 fallers (1 male; 79.4 ± 8.7 years). 13 non-fallers (4 males; 80.1 ± 6.0 years).	Subjects walked on an electronic walkway at their preferred gait speed and completed 4 trials.	Gait speed (m/s): 0.82 vs. 1.25; left and right step times (s): 0.61 vs. 0.53, 0.60 vs. 0.52; left and right heel to heel base of support (cm): 12.5 vs. 9.7, 12.4 vs. 9.6; left and right double support phase (%): 35.0 vs. 26.0, 34.0 vs. 26.0.	

Table 1. *Cont.*

Study	Inclusion and/or Exclusion Criteria Definition of Fall	Sample Characteristics	Gait Assessment	Gait Parameters Related to Falls (Fallers vs. Non-Fallers)	Gait Parameters Not Related to Falls (Fallers vs. Non-Fallers)
Wall et al., 2000 [27]	Exclusion and inclusion criteria: not reported. Definition of fall: not reported.	Community-dwelling elderly: 10 fallers (≥2 falls in past 2 years; 75.8 ± 9.3 years). 10 non-fallers (72 7 ± 4.0 years).	Gait was assessed during an expanded Timed Up and Go test. A 10 m walkway was used. A stopwatch recorded the intervals between each phase.	Front walk: gait speed (m/s): 0.81 vs. 1.23. Return walk: gait speed (m/s): 0.78 vs. 1.23.	
Hausdorff et al., 2001 [28]	Exclusion criteria: unable to follow simple instructions; nursing home residents; or life expectancy of less than 1 year. Definition of fall: not reported.	Community-dwelling elderly (16 males; 80.3 ± 5.9 years): 20 fallers (≥1 fall in 1-year follow-up). 32 non-fallers.	Subjects walked at their preferred gait speed for up to 6 min, wearing force-sensitive insoles that measured the gait rhythm on a stride-to-stride basis.	Stride time variability (SD; s): 0.11 vs. 0.05; swing time variability (SD; s): 0.04 vs. 0.03.	Gait speed (m/s): 0.71 vs. 0.91 (statistical tendency for difference, $p = 0.078$).
Kerrigan et al., 2001 [30] Kerrigan et al., 2000 [29]	Exclusion criteria: acute medical illness; diagnosis or symptoms of unstable angina or congestive heart failure; pulmonary disease diagnosis or symptoms; neurologic disorders that impair mobility; major orthopedic diagnosis in the lower back, pelvis, or lower extremities; or active joint or musculoskeletal pain. Additional exclusion criteria for fallers: falls secondary to syncope, acute illness, or other specific causes including metabolic disorders; medication side effects, true vertigo; or neurologic or lower extremity orthopedic diagnoses. Definition of fall: event that results in a subject coming to rest inadvertently on the ground or other lower level.	Community-dwelling elderly: 16 fallers (≥2 falls in last 6 months; 8 males; 77.0 ± 7.8 years). 23 non-fallers (10 males; 73.2 ± 5.5 years).	Subjects walked barefoot at their preferred and fast gait speed across a 10 m walkway. Kinematic data were collected during 3 trials using an optoelectronic motion analysis system at 100 Hz and ground reaction forces using 2 force plates.	Preferred gait speed: gait speed (m/s): 0.89 vs. 1.21; stride length (m): 0.98 vs. 1.22; hip flexion moment (Nm/kg): 0.53 vs. 0.38; hip extension moment (Nm/kg): 0.22 vs. 0.54; hip power absorption (W/kg): 0.13 vs. 0.40; hip power generation during pre-swing (W/kg): 0.43 vs. 0.92; hip adduction moment (Nm/kg): 0.47 vs. 0.57; knee flexion moment during mid-stance (Nm/kg): 0.15 vs. 0.27; knee flexion moment during pre-swing (Nm/kg): 0.07 vs. 0.24; knee power absorption during pre-swing (W/kg): 0.31 vs. 1.29; ankle power generation during pre-swing (W/kg): 1.10 vs. 1.74; hip extension (°): 11 vs. 14. Fast gait speed: gait speed (m/s): 1.34 vs. 1.57; stride length (m): 1.17 vs. 1.34; hip extension (°): 12 vs. 14.	Preferred speed: cadence (steps/min): 107 vs. 120; hip flexion moment during swing (Nm/kg): 0.08 vs. 0.11; hip power generation during loading response (W/kg): 0.50 vs. 0.50; hip abduction moment (Nm/kg): 0.08 vs. 0.07; hip internal rotation moment (Nm/kg): 0.14 vs. 0.14; hip external rotation moment (Nm/kg): 0.09 vs. 0.12; knee extension moment during terminal stance (Nm/kg): 0.16 vs. 0.16; knee power absorption during loading response (W/kg): 0.14 vs. 0.27; knee power generation during mid-stance (W/kg): 0.25 vs. 0.35; knee varus moment (Nm/kg): 0.25 vs. 0.27; knee valgus moment (Nm/kg): 0.02 vs. 0.02; knee internal rotation moment (Nm/kg): 0.14 vs. 0.13; knee external rotation moment (Nm/kg): 0.10 vs. 0.11; ankle plantarflexion moment (Nm/kg): 0.06 vs. 0.09; ankle dorsiflexion moment (Nm/kg): 0.73 vs. 0.75; ankle power absorption (W/kg): 0.43 vs. 0.44; ankle inversion moment (Nm/kg): 0.02 vs. 0.05; ankle eversion moment (Nm/kg): 0.18 vs. 0.11; ankle internal rotation

Table 1. *Cont.*

Study	Inclusion and/or Exclusion Criteria Definition of Fall	Sample Characteristics	Gait Assessment	Gait Parameters Related to Falls (Fallers vs. Non-Fallers)	Gait Parameters Not Related to Falls (Fallers vs. Non-Fallers)
					moment (Nm/kg): 0.17 vs. 0.16; ankle external rotation moment (Nm/kg): 0.09 vs. 0.09; hip flexion (°): 21 vs. 26; knee flexion during stance (°): 11 vs. 17; knee extension during stance (°): 2 vs. 2; knee flexion during swing (°): 52 vs. 58; knee extension during swing (°): 2 vs. 3; ankle plantarflexion during initial stance (°): 8 vs. 8; ankle dorsiflexion during mid-stance (°): 8 vs. 9; ankle plantarflexion (°): 14 vs. 15; ankle dorsiflexion during swing (°): 2 vs. 2; peak anterior pelvic tilt (°): 3 vs. 3. Fast gait speed: cadence (steps/min): 138 vs. 140; hip flexion (°): 25 vs. 30; knee flexion during stance (°): 16 vs. 21; knee extension during stance (°): 3 vs. 2; knee flexion during swing (°): 55 vs. 61; knee extension during swing (°): 3 vs. 6; ankle plantarflexion initial during stance (°): 8 vs. 7; ankle dorsiflexion during mid-stance (°): 6 vs. 7; ankle plantarflexion (°): 14 vs. 16; ankle dorsiflexion during swing (°): 1 vs. 2; peak anterior pelvic tilt (°): 4 vs. 4.
Kemoun et al., 2002 [31]	Exclusion and inclusion criteria: not reported. Definition of fall: unexpected event when a subject falls to the ground from the same or an upper level (including falls on stairs and onto a piece of furniture).	Community-dwelling elderly (66.7 ± 4.8 years): 16 fallers (≥1 fall in 1-year follow-up; 12 males). 38 non-fallers (26 males).	Subjects walked barefoot at their preferred gait speed across a 10 m walkway. Kinematic data were collected during 5 trials using a 5-camera optoelectronic motion analysis system at 50 Hz and ground reaction force using two integrated force platforms at 250 Hz.	Gait speed (m/s): 0.96 vs. 1.29; double support phase (%): 27.8 vs. 23.2; ankle moment peak (Nm/kg): 25 vs. 23; ankle plantarflexion during second double support (°): 19 vs. 23; ankle dorsiflexion at beginning of swing (°): 7 vs. 13; hip power variation (W/kg): 1.02 vs. 2.04; hip moment peak (Nm/kg): −0.54 vs. −0.97; hip moment variation (Nm/kg): 0.88 vs. 1.60; hip displacement (°): 40 vs. 47.	Cadence (step/min): 99 vs. 108 (statistical tendency for difference, $p = 0.059$); stride length (m): 1.12 vs. 1.31; step length (m): 0.57 vs. 0.65; stride time (s): 1.20 vs. 1.11 (statistical tendency for difference, $p = 0.058$); single support phase (%): 37 vs. 38.2; step time (%): 49.3 vs. 50.0; single support start (%): 13.5 vs. 13.6; double support start (%): 50.7 vs. 50.0; swing start (%): 64.6 vs. 62.1 (statistical tendency for difference, $p = 0.077$); ankle power peak (W/kg): 2.5 vs. 3.1; ankle moment peak (Nm/kg): 1.58 vs. 1.54; knee power peak (W/kg): −0.81 vs. −1.35; knee power variation (W/kg): 0.91 vs. 1.42; knee moment peak (Nm/kg): −0.17 vs. −0.04;

Table 1. *Cont.*

Study	Inclusion and/or Exclusion Criteria Definition of Fall	Sample Characteristics	Gait Assessment	Gait Parameters Related to Falls (Fallers vs. Non-Fallers)	Gait Parameters Not Related to Falls (Fallers vs. Non-Fallers)
					knee moment variation (Nm/kg): 0.74 vs. 0.74; knee position peak (°): 62 vs. 63; knee displacement (°): 50 vs. 52; hip power peak (W/kg): −0.93 vs. −1.23; hip position peak (°): 57 vs. 61.
Auvinet et al., 2003 [32]	Inclusion criteria (fallers): recently hospitalized due to falls; living at home; and no pelvic or leg length asymmetries. Inclusion criteria (non-fallers): no history of musculoskeletal, neurological, or gait disorder; living at home; and no marked pelvic asymmetry or leg length differences. Definition of fall: not reported.	Community-dwelling elderly: 20 fallers (≥1 fall in past year; 2 males; 80.7 ± 5.2 years). 33 non-fallers (18 males; 77.2 ± 6.5 years).	Subjects walked at their preferred gait speed across a 10 m walkway using their own shoes. Gait parameters were collected using an accelerometer sensor system (50 Hz).	Gait speed (m/s): 0.73 vs. 1.24; stride length (m): 0.86 vs. 1.28; stride frequency-cadence (Hz): 0.86 vs. 0.97; stride symmetry: 173 vs. 211.	
Mbourou et al., 2003 [33]	Exclusion criteria: Parkinson's disease or Alzheimer's disease. Inclusion criteria (fallers): living in a nursing home. Definition of fall: not reported.	Elderly: 9 fallers (≥1 fall in past year; mean age 80.0 years, range 74.0–91.0). 8 non-fallers (mean age 73.0 years, range 65.0–82.0).	Subjects were asked to initiate gait and walk at least 3 strides. The length of the first step and subsequent strides were collected using transducers. Gait parameters were derived from the displacement signal obtained from each foot. More than 10 trials were collected.	First step length (m): 0.30 vs. 0.53; first step length variability (SD; m): 0.13 vs. 0.06; first double support phase (%): 32 vs. 22; second stride length (m): 0.68 vs. 0.92; second stride length variability (SD; m): 0.10 vs. 0.05; double support phase for subsequent strides (%): 37 vs. 32.	
Pijnappels et al., 2005 [103]	Exclusion and inclusion criteria: not reported. Definition of fall: when the vertical force in the ropes exceeded 200 N during trials when one obstacle appeared from the ground unexpectedly to catch the subject's swing limb.	Community-dwelling elderly: 7 fallers (when the vertical force in the ropes exceeded 200 N during trip trials; 1 male; 67.9 ± 2.6 years). 4 non-fallers (3 males; 66.5 ± 3.3 years).	Subjects walked at preferred gait speed over a platform. Kinematic and ground reaction force data were collected using a 4-camera optoelectronic motion analysis system and a force plate (100 Hz).		Stance time; double support time.
Chiba et al., 2005 [34]	Exclusion criteria: Mini Mental Status Examination score < 24; arthritis in lower limbs; back, knee, or hip chronic pain; Parkinson's disease; Ménière's syndrome; cerebellar signs; or peripheral neuropathy under standard neuropsychological assessment. Inclusion criteria: being medically stable;	Community-dwelling elderly: 25 fallers (≥2 falls in past year; 11 males; 76.0 ± 6.6 years). 31 non-fallers (11 males; 74.9 ± 7.2 years).	Subjects walked barefoot or with their own shoes on a 6 m walkway. During 2 continuous trials, kinematic data were collected using a 5-camera optoelectronic motion analysis system (60 Hz). MTC was analyzed.	Gait speed (m/s): 0.66 vs. 0.99; stride length (m): 0.77 vs. 1.06; stride time (s): 1.20 vs. 1.08; MTC (mm): 12.0 vs. 15.2; MTC variability (CV; %): 29 vs. 25; maximum foot angle with ground (°): 7.4 vs. 14.3; variability in the maximum foot angle with ground (CV; %):	

Table 1. *Cont.*

Study	Inclusion and/or Exclusion Criteria Definition of Fall	Sample Characteristics	Gait Assessment	Gait Parameters Related to Falls (Fallers vs. Non-Fallers)	Gait Parameters Not Related to Falls (Fallers vs. Non-Fallers)
	comprehending the nature of the study and our instructions; and being able to stand up and walk independently without an assistance device. Definition of fall: a sudden unintentional change in position causing a subject to land at a lower level or ground.			34 vs. 19; maximal ML displacement of trunk center (mm): 0.23 vs. 0.18; variability in the maximal ML displacement of trunk center (CV; %): 6 vs. 1.	
Barak et al., 2006 [35]	Exclusion criteria: cardiopulmonary, musculoskeletal, somatosensory, or neurological disorders or severe visual and vestibular loss. Definition of fall: not reported.	Community-dwelling elderly: 21 fallers (≥1 fall in last 6 months; 8 males; 73.8 ± 6.4 years). 27 non-fallers (14 males; 72.1 ± 4.9 years).	Subjects walked at their preferred gait speed; treadmill speed was gradually increased from 0.18 m/s to 1.52 m/s in steps of 0.225 m/s and then decreased. During 1 min in each step, kinematic data were collected using an optoelectronic motion analysis system (100 Hz). CoM was calculated.	Effects of gait speed: cadence: in all gait speeds; stride length: only in 1.3 m/s gait speed; CoM lateral sway: from 1.07 m/s gait speed; ankle plantarflexion: from 1.07 m/s gait speed; hip extension: from 0.85 m/s gait speed; hip flexion: from 0.85 m/s gait speed; hip flexion variability: only in 1.52 m/s gait speed. Effects of stride frequency: stride length: in 1.1 and 1.2 stride frequencies; lateral body sway: in 0.6–1.1 stride frequencies; hip extension: in 0.5–1.0 stride frequencies; hip extension variability: in 0.9–1.2 stride frequencies; hip flexion: in 1.0 stride frequency; hip flexion variability: in 0.9–1.2 stride frequency.	Effects of stride frequency: ankle plantarflexion.
Toulotte et al., 2006 [36]	Exclusion criteria: lower limb fracture or surgery; use of walking aid or foot orthosis; cognitive disorders; auditory, ocular, or vestibular problems; head trauma with/without loss of consciousness; stroke; carpal tunnel syndrome; or sores on lower limbs or corns. Inclusion criteria: ≥60 years and stable medical treatment (for at least 3 months). Definition of fall: any event that led to an unplanned contact with a supporting surface.	Community-dwelling elderly women: 21 fallers (≥1 fall in past 2 years; 70.4 ± 6.4 years). 19 non-fallers (67.0 ± 4.8 years).	Subjects walked at preferred gait speed across a 10 m walkway. During 10 trials, kinematic data were collected using an optoelectronic motion analysis system (50 Hz) and three force plates (250 Hz).		Cadence (steps/min): 116 vs. 119; gait speed (m/s): 1.08 vs. 1.12; stride time (s): 1.04 vs. 1.02; step time (s): 0.53 vs. 0.51; single support time (s): 0.48 vs. 0.48; stride length (m): 1.13 vs. 1.09; step length (m): 0.56 vs. 0.60.

Table 1. *Cont.*

Study	Inclusion and/or Exclusion Criteria Definition of Fall	Sample Characteristics	Gait Assessment	Gait Parameters Related to Falls (Fallers vs. Non-Fallers)	Gait Parameters Not Related to Falls (Fallers vs. Non-Fallers)
Karmakar et al., 2012 [38] Karmakar et al., 2007 [37]	Exclusion and inclusion criteria: not reported. Definition of fall: not reported.	Community-dwelling elderly women: 10 fallers (≥1 fall in past year; 72.2 ± 3.1 years). 27 non-fallers (69.1 ± 5.1 years).	Subjects walked on a treadmill at preferred gait speed. During the first 500 continuous gait cycles, MTC data were collected using a 2D motion analysis system and analyzed by an ApEn and a SampEn. ApEn was calculated with m = 3 and r from 0 to 90% of the calculated SD. SampEn was calculated with m varying from 2 to 4 and r from 0 to 90% of the calculated SD.	Gait speed (m/s): 0.91 vs. 1.29; MTC (cm): 2.02 vs. 1.25; MTC variability (SD; cm): 0.47 vs. 0.32. For r < 0.26 * SD, the mean MTC ApEn of fallers was higher than non-fallers. For r ≥ 0.26 * SD, the mean MTC ApEn of fallers was smaller than non-fallers. MTC SampEn values of fallers were lower compared to non-fallers for all m and r.	
Newstead et al., 2007 [39]	Exclusion criteria: neurological or orthopedic conditions. Inclusion criteria: ≥60 years; be able to walk 1 mile nonstop; and free of neurological or orthopedic impairments. Definition of fall: not reported.	Community-dwelling elderly: 18 fallers (≥1 fall in past year; 3 males; 78.1 ± 7.2 years). 30 non-fallers (6 males; 75.8 ± 5.1 years).	Subjects walked 5–7 trials at three different gait speeds (slow, preferred, and fast) across a 10 m walkway using laced walking shoes. Spatiotemporal data were collected using a 6-camera optoelectronic motion analysis system (60 Hz) and four force plates (250 Hz).	Preferred gait speed: gait speed (lower fallers); cadence (lower fallers); step length (lower fallers); stride length (lower fallers); single support time (lower fallers); double support time (higher fallers). Fast gait speed: stride length (lower fallers); cadence (lower fallers); gait speed (lower fallers).	Slow gait speed: gait speed; cadence; step length; stride length; single support time; double support time. Fast gait speed: step length; single support time; double support time.
Barrett et al., 2008 [40]	Exclusion criteria: limited pulmonary and cardiac function; use of pacemakers; or cognitive impairment. Inclusion criteria: independent ambulation for at least 6 m. Definition of fall: subject who had fallen from vertical to horizontal.	Community-dwelling elderly: 9 fallers (≥2 falls in past year; 5 males; 76.0 ± 5.0 years). 10 non-fallers (5 males 69.0 ± 5.0 years).	Subjects walked at preferred gait speed over a 6 m walkway. Gait events were detected using footswitches embedded in the left shoe. Three walks were recorded for analysis.	Stance time (higher fallers); stride time (higher fallers); stance phase (higher fallers); stance time variability (SD; higher fallers); stride time variability (SD; higher fallers).	Swing time; swing time variability (SD).
Khandoker et al., 2008 [41]	Exclusion and inclusion criteria: not reported. Definition of fall: not reported.	Community-dwelling elderly: 10 fallers (≥1 fall; 72.2 ± 3.1 years). 14 non-fallers (71.0 ± 2.1 years).	Subjects walked 10–20 min on the treadmill. MTC data were collected using a 2D motion analysis system (50 Hz) and analyzed by ApEn and Poincaré Plot Indexes. ApEn was calculated with m = 3 and r from 0 to 90%. SampEn was calculated with m varying from 2 to 3 and r from 15%.	MTC ApEn values: 0.18 vs. 0.13; MFC variability (SD; cm): 0.48 vs. 0.35; Poincaré width: 0.72 vs. 0.51; Poincaré length: 1.15 vs. 0.89.	MFC (cm): 2.01 vs. 1.65; Poincaré width/Poincaré length: 0.64 vs. 0.64.

Table 1. *Cont.*

Study	Inclusion and/or Exclusion Criteria Definition of Fall	Sample Characteristics	Gait Assessment	Gait Parameters Related to Falls (Fallers vs. Non-Fallers)	Gait Parameters Not Related to Falls (Fallers vs. Non-Fallers)
Khandoker et al., 2008 [42]	Exclusion and inclusion criteria: not reported. Definition of fall: not reported.	Community-dwelling elderly women: 10 fallers (≥1 fall in past year; 72.2 ± 3.1 years). 27 non-fallers (69.1 ± 5.1 years).	Subjects walked 10 min on a treadmill at preferred gait speed. MTC data were collected using a 2D motion analysis system (50 Hz). The following variability indices were quantified: Poincaré plot indices (SD1, SD2, SD1/SD2); wavelet-based multiscale exponent; and detrended fluctuation analysis exponent to investigate the presence of long-range correlations in MTC time series.	MTC (cm): 2.02 vs. 1.25. Wavelet-based multiscale exponent, SD1/SD2, and SD2 of critical MTC parameters were found to be potential markers to be able to reliably identify fallers from non-fallers.	
Lockhart & Liu, 2008 [43]	Exclusion and inclusion criteria: not reported. Definition of fall: not reported.	Community-dwelling elderly: 4 fallers (≥1 fall in past 6 months; 70.1 ± 3.0 years). 4 non-fallers (71.3 ± 6.5 years).	Subjects walked for 1 min on a treadmill at their preferred gait speed. One dual-axial accelerometer was placed on the right anterior superior iliac spine (125 Hz). Maximum Lyapunov exponent was used to analyze these data. Two infrared-reflective markers were placed bilaterally on the heels for kinematic capture with a 6-camera optoelectronic motion analysis system (120 Hz).	Maximum Lyapunov exponent: 2.39 vs. 1.99; step length (m): 0.33 vs. 0.60; gait speed (m/s): 0.57 vs. 1.16.	Heel contact velocity (m/s): 0.32 vs. 0.43; step duration (s): 1.19 vs. 1.04.
Verghese et al., 2009 [14]	Exclusion criteria: severe audiovisual loss; bed-bound due to illness; or institutionalization. Definition of fall: unintentionally coming down on the floor or to a lower level, not due to a major intrinsic or extrinsic event.	Community-dwelling elderly (227 males; mean age 80.6 years): 226 fallers (115 fell once and 111 had recurrent falls; mean age 81.1 years). 371 non-fallers (mean age 80.1 years).	Subjects walked at preferred gait speed during 2 trials on a computerized walkway with embedded pressure sensors using comfortable footwear. Generalized estimating equations with a binomial distribution to model the probability of fall.	Slower gait speed (risk ratio per 10 cm/s decrease 1.069, 95% confidence interval: 1.001–1.142) is associated with a higher risk of falls. Predicted fall risk: swing phase (RR 1.406, 95% confidence interval 1.027–1.926); double-support phase (RR 1.165, 95% confidence interval: 1.026–1.321); swing time variability–CV (RR 1.007, 95% confidence interval: 1.004–1.010); stride length variability–CV (RR 1.076, 95% confidence interval: 1.030–1.111).	
Greany & Di Fabio, 2010 [44]	Inclusion criteria: ≥70 years; living at home; can walk at least 30 feet without stopping; Mini Mental Status Examination score > 23; corrected visual acuity of at least 20/70; and peripheral visual field of 30°. Definition of fall: unintentionally coming to rest on the ground.	Community-dwelling elderly: 12 fallers (≥1 fall in past year; 3 males, 86.0 ± 4.8 years). 21 non-fallers (7 males, 81.0 ± 5.0 years).	Subjects walked quickly but safely along a walkway of four irregularly spaced stepping targets. During 6 trials, a video-based motion analysis system was used for collecting kinematic data.		Maximum foot AP velocity (m/s): 1.91 vs. 2.05; maximum foot vertical velocity (m/s): 0.54 vs. 0.53; average foot AP velocity (m/s): 0.36 vs. 0.36; step time (s): 1.64 vs. 1.51; step length (m): 0.76 vs. 0.83; swing time (s): 0.67 vs. 0.68; double support time (s): 0.45 vs. 0.35.

Table 1. *Cont.*

Study	Inclusion and/or Exclusion Criteria Definition of Fall	Sample Characteristics	Gait Assessment	Gait Parameters Related to Falls (Fallers vs. Non-Fallers)	Gait Parameters Not Related to Falls (Fallers vs. Non-Fallers)
Greene et al., 2010 [45]	Inclusion criteria: ≥60 years; able to walk independently with or without help; and able to provide informed consent. Definition of fall: unexpected loss of stability resulting in coming to rest on the floor or an object below the knee level.	Community-dwelling elderly: 207 fallers (≥1 fall in past 5 years; 44 males; 74.0 ± 7.3 years). 142 non-fallers (59 males; 71.1 ± 6.9 years).	Gait was assessed during the Timed Up and Go test through two wearable tri-axial accelerometer sensors placed on each shank.	Cadence (steps/min): 99 vs. 108; double support (s): 0.4 vs. 0.5; step time (s): 0.7 vs. 0.6; minimum shank ML angular velocity (lower fallers); mean shank ML angular velocity (lower fallers); maximum shank ML angular velocity (lower fallers); minimum shank AP angular velocity (lower fallers); mean shank AP angular velocity (lower fallers); maximum shank AP angular velocity (lower fallers).	Stance time (s): 0.8 vs. 0.8; single support time (s): 0.8 vs. 0.8; stride time (s): 1.2 vs. 1.2; swing time (s): 0.5 vs. 0.5; single support time variability (CV; %): 22.9 vs. 21.1; double support variability (CV; %): 80.7 vs. 82.6; swing time variability (CV; s): 28.1 vs. 31.0; stride time variability (CV; s): 24.0 vs. 23.4; step time variability (CV; s): 42.0 vs. 40.3; stance time variability (CV; s): 43.3 vs. 45.0.
Mickle et al., 2010 [46]	Inclusion criteria: ≥60 years; living independently in the community; passed the Short Portable Mental Status Questionnaire; able to ambulate for at least 10 m with or without aid; no neurological diseases; and own transport to a testing venue in the community. Definition of fall: unintentionally coming to rest on the ground or other lower level, not as a result of a major intrinsic event (e.g., stroke).	Community-dwelling elderly: 107 fallers (≥1 fall in 1-year follow-up; 49 males; 71.6 years). 196 non-fallers (105 males; 71.2 years).	Five trials were recorded with a two-step gait initiation protocol at a preferred walking speed. A pressure platform was used.	Peak pressure (KPa): 776 vs. 699; pressure–time integral (KPa): 349 vs. 311.	
Bhatt et al., 2011 [104]	Exclusion criteria: Folstein Mini Mental Status Examination score < 25 or classified as osteopenic or osteoporotic. Definition of fall: if the force recorded on the safety harness load cell force exceeded 30% of the body weight.	Community-dwelling elderly (44 males): 59 fallers (the force recorded on the safety harness load cell force exceeded 30% of the body weight; 71.6 ± 4.6 years). 56 recoveries (71.4 ± 5.1 years).	Subjects walked for 9–12 trials on a 7 m walkway using their own athletic shoes. Kinematic and ground reaction force data were collected during 5 trials using an 8-camera optoelectronic motion analysis system (120 Hz) and one force platform (600 Hz).		Absolute CoM velocity (m/s): 0.95 vs. 1.0; step length (m): 0.34 vs. 0.34.
Kirkwood et al., 2011 [47]	Exclusion criteria: orthopedic or neurological diseases that could affect gait performance. Inclusion criteria: female; ≥60 years; and ability to walk without assistance. Definition of fall: unexpected event in which a subject comes to rest on a lower level.	Community-dwelling elderly women: 45 fallers (≥2 falls in past 12 months; 74.0 ± 5.6 years). 44 non-fallers (absence of falls or 1 fall in past 12 months; 70.7 ± 5.4 years).	Subjects walked barefoot on a 6 m rubber mat while EMG recorded soleus, tibialis anterior, and gastrocnemius muscle signals. Footswitches tracked gait events.	Gastrocnemius activity during stance phase (%): 16.9 vs. 19.8; stride time (s): 1.2 vs. 1.3.	Tibialis anterior activity during stance phase (%): 9.2 vs. 9.3; soleus activity during stance phase (%): 22.1 vs. 24.9; gastrocnemius latency activity (s): 0.30 vs. 0.30; tibialis anterior latency activity (s): 0.04 vs. 0.03; soleus latency activity (s): 0.30 vs. 0.30; swing phase (%): 40.2 vs. 39.7; stance phase (%): 59.8 vs. 60.3.

Table 1. *Cont.*

Study	Inclusion and/or Exclusion Criteria Definition of Fall	Sample Characteristics	Gait Assessment	Gait Parameters Related to Falls (Fallers vs. Non-Fallers)	Gait Parameters Not Related to Falls (Fallers vs. Non-Fallers)
Lázaro et al., 2011 [48]	Exclusion criteria: ≥65 years; severe cognitive deterioration; unable to stand; or terminally ill. Inclusion criteria (fallers): had visited their General Practitioner or Geriatrician due to the occurrence of falls. Definition of fall: not reported.	Community-dwelling elderly: 99 fallers (≥2 falls in past 6 months; 17 males; 78.0 ± 5.0 years). 113 non-fallers (no falls in past 6 months).	Gait assessments of subjects were evaluated using the Walk Across test.	Gait speed (m/s): 0.34 vs. 0.50.	
Lugade et al., 2011 [49]	Inclusion criteria: no history of head trauma, neurological or heart diseases; muscle, joint, or orthopedic disorder; visual impairment that was uncorrected by glasses; persistent vertigo; or lightheadedness. Definition of fall: not reported.	Community-dwelling elderly: 10 fallers (≥1 fall in past year; 78.9 ± 4.9 years). 10 non-fallers (75.4 ± 7.0 years).	Subjects walked barefoot at preferred gait speed along a 10 m walkway. Kinematic and ground reaction force data were collected during 5 trials using an 8-camera optoelectronic motion analysis system (60 Hz) and two integrated force platforms. CoM and CoP data were calculated.	Gait speed (m/s): 1.02 vs. 1.26; CoM–CoP AP distance at heel strike (cm): 41.6 vs. 52.4.	At heel strike (CoM inside base of support): CoM stability margin (cm): 3.9 vs. 3.5; distance to centroid (cm): 2.5 vs. 2.2; interaction of the CoM position and velocity distance to border (cm): 17.5 vs. 18.7; time to contact (s): 0.17 vs. 0.15; base of support (m^2): 0.40 vs. 0.43. At toe-off (CoM outside base of support): CoM separation (cm): 8.3 vs. 10.4; distance to centroid (cm): 21.4 vs. 23.4; time to contact (s): 0.12 vs. 0.11; base of support area (m^2): 0.23 vs. 0.22; CoM–CoP ML distance at heel strike (cm): 6.6 vs. 7.3.
Panzer et al., 2011 [50]	Exclusion criteria: Mini Mental Status Examination score < 24; body mass index ≥30 kg/m^2; blindness; neurologic, orthopedic, or visual disorders that impair mobility; or non-English speaking. Definition of fall: not reported.	Community-dwelling elderly: 47 fallers (≥2 non-injury falls or ≥1 injury fall in past year; 80.1 ± 6.2 years). 27 non-fallers (75.1 ± 6.5 years).	Two self-paced out and back walks (8.1 m) were performed; average gait speed was calculated, and the fastest performance was used.	Average gait speed (m/s): 0.64 vs. 0.86.	
Scanaill et al., 2011 [51]	Exclusion and inclusion criteria: not reported. Definition of fall: not reported.	Community-dwelling elderly: 182 fallers (>1 fall in past year or ≥1 fall that resulted in a loss of consciousness, a fractured bone, or severe injury in past year; 40 males; 74.5 ± 7.2 years). 139 non-fallers (60 males; 70.3 ± 6.8 years).	Subjects walked at a preferred gait speed along a 6 m pressure-sensing walkway. Two kinematic sensors were worn on the subject's shanks.	Stride length (m): 1.08 vs. 1.23; stride width (m): 0.12 vs. 0.11; step length (m): 0.54 vs. 0.61; step width (m): 0.56 vs. 0.63.	Stride time (s): 1.23 vs. 1.20; stance time (s): 0.81 vs. 0.79; swing time (s): 0.51 vs. 0.50; step time (s): 0.66 vs. 0.66; single support (%): 75.9 vs. 78.2; double support (%): 34.6 vs. 34.3. Variability (CV): stride length (%): 8.6 vs. 7.8; stride width (%): 25.0 vs. 25.3; step length (%): 14.1 vs. 12.7; step width (%): 12.4 vs. 10.6; base width (%): 24.7 vs. 25.3; stride time (%): 19.2 vs. 18.6; stance time (%): 30.3 vs. 33.0; swing time (%): 32.4 vs. 31.0; step time (%): 34.2 vs. 31.8; single support (%): 21.4 vs. 20.1; double support (%): 61.4 vs. 62.6.

Table 1. *Cont.*

Study	Inclusion and/or Exclusion Criteria Definition of Fall	Sample Characteristics	Gait Assessment	Gait Parameters Related to Falls (Fallers vs. Non-Fallers)	Gait Parameters Not Related to Falls (Fallers vs. Non-Fallers)
Uemura et al., 2012 [52]	Exclusion criteria: severe cardiac, pulmonary, or musculoskeletal disorders; diseases associated with a high risk of falling; inability to execute arithmetic tasks; serious visual impairment not correctable with spectacles; or inability to follow multiple commands. Inclusion criteria: ≥65 years; minimal hearing and visual impairments; and ability to ambulate independently. Definition of fall: an event where a subject unintentionally comes to rest on the ground or another lower level; falls resulting from extraordinary environmental factors were excluded.	Community-dwelling elderly (65–93 years): 22 fallers (≥1 fall in past year). 35 non-fallers.	Subjects walked along a 2 m walkway as quickly as possible after a visual cue. CoP data were collected by force plate during 3 trials. Step initiation—first ML deviation of CoP towards swing leg. Reaction phase—time from cue to step initiation. Anticipatory postural adjustment phase—time from step initiation to foot-off.		Reaction phase (s): 0.31 vs. 0.29; anticipatory postural adjustment phase (s): 0.46 vs. 0.44.
Chen & Chou, 2013 [53]	Inclusion criteria: walk without an assistive device; no history of neurological or musculoskeletal deficits (e.g., amputation, cerebral vascular accident, significant head trauma, or Parkinson's disease); and no uncorrectable visual impairment, vestibular dysfunction, or dementia. Definition of fall: unexpected event where the subject falls to the ground from an upper level; falls caused by syncope or major intrinsic events were excluded.	Community-dwelling elderly: 15 fallers (≥2 falls in past year; 4 males; 77.7 ± 7.7 years). 15 non-fallers (6 males; 76.2 ± 4.2 years).	Subjects performed the Timed Up and Go test while barefoot. Kinematic and CoM data and ground reaction force were collected during 4 trials using a 10-camera optoelectronic motion analysis system (600 Hz) and one force platform (960 Hz).	Step length (m): 0.42 vs. 0.52; CoM AP velocity at stance-off (lower fallers); AP inclination of CoM–ankle at stance-off (°): −2.4 vs. −6.8; total CoM kinetic energy at swing-off (J): 6.6 vs. 10.4; total CoM kinetic energy at stance-off (J): 20.6 vs. 31.9.	Step width (m): 0.23 vs. 0.21; CoM AP velocity at swing-off; AP inclination of CoM–ankle at swing-off (°): 7.7 vs. 6.9.
Chen et al., 2013 [54]	Inclusion criteria: walk without an assistive device; no history of neurological or musculoskeletal deficits (e.g., amputation, cerebral vascular accident, significant head trauma, or Parkinson's disease); and no uncorrectable visual impairment, vestibular dysfunction, or dementia. Definition of fall: unexpected event where the subject falls to the ground from an upper level; falls caused by syncope or major intrinsic events were excluded.	Community-dwelling elderly: 10 fallers (≥2 falls in past year; 2 males; 75.9 ± 4.1 years). 10 non-fallers (3 males; 75.5 ± 3 0 years).	Subjects performed the Timed Up and Go test while barefoot. Kinematic and CoM data and ground reaction force were collected during 4 trials using a 10-camera optoelectronic motion analysis system (600 Hz) and one force platform (960 Hz).	Braking force (N/kg): −0.83 vs. −0.43 propulsive force (N/kg): 3.48 vs. 5.04; ankle moment at swing-off: 0.11 vs. −0.03.	Trunk angle (°): 32.9 vs. 31.4; hip moment at swing-off (Nm/kg): 0.45 vs. 0.48; knee moment at swing-off (Nm/kg): 0.42 vs. 0.54.

Table 1. *Cont.*

Study	Inclusion and/or Exclusion Criteria Definition of Fall	Sample Characteristics	Gait Assessment	Gait Parameters Related to Falls (Fallers vs. Non-Fallers)	Gait Parameters Not Related to Falls (Fallers vs. Non-Fallers)
Chiu & Chou, 2013 [55]	Inclusion criteria: no current histories of neurological or musculoskeletal deficits that affect walking and no uncorrectable visual impairment, vestibular dysfunction, dementia, or depression. Definition of fall: not reported.	Community-dwelling elderly: 15 fallers (≥2 falls in past year; 3 males; 72.9 ± 4.1 years). 15 non-fallers (8 males; 75.7 ± 4.7 years).	Subjects walked barefoot along a 10 m walkway at preferred gait speed. Kinematic data were collected during 5 trials using a 10-camera optoelectronic motion analysis system (60 Hz). SD is used to analyze variability.	Gait speed (m/s): 1.07 vs. 1.22; stance phase (%): 62.6 vs. 60.9; swing phase (%): 37.4 vs. 39.1; single support (%): 37.4 vs. 39.0; double support (%): 25.2 vs. 21.9. Variability in inter-joint coordination during stance phase (SD): knee–ankle (higher fallers); ankle (higher fallers). Variability inter-joint coordination during swing phase (SD): knee–ankle (higher fallers).	Cadence (steps/min): 115 vs. 116. Variability inter-joint coordination during stance phase (SD): hip; knee; hip–knee. Variability inter-joint coordination during swing phase (SD): hip; knee; ankle; hip–knee.
Fritz et al., 2013 [56]	Exclusion criteria: orthopedic or neurologic conditions that altered walking. Inclusion criteria: capable of walking unassisted for more than 10 feet and understanding the study's objective. Definition of fall: not reported.	Community-dwelling elderly: 12 fallers (≥1 fall in past 6 months; 86.3 ± 4.7 years). 50 non-fallers (85.4 ± 7.1 years).	Subjects walked during 3 trials at a preferred gait speed along a 6 m pressure-sensing walkway.	Gait speed (m/s): 0.89 vs. 1.0; stride length (m): 0.85 vs. 1.02.	Base of support (cm): 12.3 vs. 10.2; swing phase (%): 31.5 vs. 33.4; stance phase (%): 68.4 vs. 66.6; double support (%): 37 vs. 33; step time variability (CV): 7.3 vs. 6.4.
Weiss et al., 2013 [57]	Exclusion criteria: previously clinically diagnosed with any gait or balance disorders and Mini Mental Status Examination score < 24. Definition of fall: any stability disturbance that caused significant contact with the floor.	Community-dwelling elderly: 32 fallers (≥2 fall in past year; 35% males; 77.9 ± 5.1 years). 39 non-fallers (<2 fall in past year; 36% males; 78.8 ± 4.4 years).	Subjects walked for 1 min at preferred gait speed (laboratory assessments). A portable tri-axial accelerometer sensor (100 Hz) was worn on the lower back. Subjects also wore a portable accelerometer sensor (100 Hz) for 3 days.	Laboratory assessment: gait speed (m/s): 0.97 vs. 1.19; step duration (s): 0.55 vs. 0.52. 3-day assessment: step time; stride time. Fallers presented higher variability in the lower back vertical axis and lower variability in the lower back ML axis.	
Marques et al., 2013 [59] Marques et al., 2013 [58]	Exclusion criteria: Mini Mental Status Examination score < 20; cardiovascular disease; Berg balance scale score < 36; hemiparesis; pain of the lower limbs or trunk; or progressive motor disorder. Definition of fall: any stability disturbance that caused significant contact with the floor.	Community-dwelling elderly women: 15 fallers (≥1 fall in past year; 69.6 ± 8.0 years). 22 non-fallers (66.1 ± 6.2 years).	Subjects walked at preferred gait speed for 1 min on a walkway and for 10 min on a treadmill. Gait kinematic parameters and EMG activity were assessed using a 7-camera optoelectronic motion analysis system (100 Hz) and an 8-channel telemetry EMG system (2000 Hz).	Hip position at toe-off (°): 9.5 vs. 5.4; muscle activation at initial stance: biceps femoris (%): 36.4 vs. 24.1; muscle activation at final stance: gluteus maximus (%): 86.4 vs. 52.3; muscle activation before heel contact: internal oblique (%): 8.3 vs. 15.7; biceps femoris (%): 45.5 vs. 31.3.	Gait speed on walkway (m/s): 1.1 vs. 1.3; gait speed on treadmill (m/s): 0.9 vs. 0.9; step time (s): 0.23 vs. 0.26; step length (m): 0.51 vs. 0.50; step width (m): 0.14 vs. 0.17; ankle angular position at heel contact (°): 6.4 vs. 5.9. Muscle activation at initial stance: internal oblique (%): 97.2 vs. 100.3; rectus femoris (%): 143.8 vs. 130.6; tibialis anterior (%): 106.7 vs. 122.8; multifidus (%): 150.5 vs. 147.7, gluteus maximus (%): 154.7 vs. 179.9. Muscle activation at final stance: internal oblique (%): 117.5 vs. 105.1; rectus femoris (%): 89.5 vs. 80.1,

Table 1. *Cont.*

Study	Inclusion and/or Exclusion Criteria Definition of Fall	Sample Characteristics	Gait Assessment	Gait Parameters Related to Falls (Fallers vs. Non-Fallers)	Gait Parameters Not Related to Falls (Fallers vs. Non-Fallers)
					multifidus (%): 76.1 vs. 82.8; biceps femoris (%): 43.8 vs. 50.1; gastrocnemius lateralis (%): 91.7 vs. 75.8. Muscle activation before heel contact: rectus femoris (%): 12.7 vs. 15.2; tibialis anterior (%): 40.0 vs. 30.1; multifidus (%): 16.4 vs. 18.4; gluteus maximus (%): 12.0 vs. 16.8; gastrocnemius lateralis (%): 7.2 vs. 14.9. Muscle activation after toe-off: internal oblique (%): 21.6 vs. 20.5; rectus femoris (%): 10.9 vs. 15.9; tibialis anterior (%): 35.1 vs. 31.9; gluteus maximus (%): 6.8 vs. 10.1; biceps femoris (%): 16.2 vs. 13.1; gastrocnemius lateralis (%): 7.9 vs. 12.2.
Ayoubi et al., 2014 [60]	Exclusion criteria: <65 years; institutionalization; non–French-speaking; acute medical illness during the past month; diagnosis of dementia; score > 2 on item 22 of Unified Parkinson's Disease Rating Scale; severe orthopedic diagnoses of lumbar vertebra, pelvis, or lower extremities; or inability to walk 6 m unassisted. Definition of fall: subject unintentionally coming to rest on the ground or other lower level, and not as the result of a major intrinsic event.	Community-dwelling elderly: 109 fallers with a fear of falling (24 males; 71 ± 5.2 years). 101 fallers with no fear of falling (29 males; 70.8 ± 5.5 years). 194 non-fallers with fear of falling (83 males; 70.5 ± 5.0 years). 619 non-fallers with no fear of falling (368 males; 70.3 ± 4.8 years).	Subjects walked 1 trial at their preferred gait speed along a 6 m pressure-sensing walkway.	Fallers with fear of falling vs. non-fallers with no fear of falling: gait speed (m/s): 0.96 vs. 1.11; stride time variability (CV; %): 3.0 vs. 2.0.	Fallers with no fear of falling vs. non-fallers with no fear of falling: gait speed (m/s) 1.07 vs. 1.11; stride time variability (CV; %): 2.0 vs. 2.0. Fallers with no fear of falling vs. non-fallers with fear of falling: gait speed (m/s) 1.07 vs. 1.03; stride time variability (CV; %): 2.0 vs. 3.0. Fallers with fear of falling vs. non-fallers with fear of falling: gait speed (m/s) 0.96 vs. 1.03; stride time variability (CV; %): 3.0 vs. 3.0.
Barelle et al., 2014 [61]	Exclusion criteria: vascular stroke with motor or sensory after-effects; Parkinson's disease; hip or knee prosthesis; or fracture of leg or ankle which would have impaired gait. Definition of fall: not reported.	Community-dwelling elderly: 6 fallers (≥1 fall in past 6 months; 68.0 ± 4.0 years). 6 non-fallers (2 males; 69.0 ± 3.0 years).	Subjects walked at preferred gait speed on a 10 m walkway. Gait kinematic parameters were assessed using an 8-camera optoelectronic motion analysis system (100 Hz).		Stride and step length (m): 1.13 vs. 1.18; 0.57 vs. 0.59; stride length (% height): 70 vs. 74; cadence (strides/s): 0.87 vs. 0.92; cadence (steps/s): 1.73 vs. 1.84; step length (% height): 35 vs. 37; cycle time (s): 1.17 vs. 1.09; gait speed (m/s): 0.99 vs. 1.08; hip, knee, ankle displacements (°): 21 vs. 21; 58 vs. 58; 38 vs. 37.

Table 1. *Cont.*

Study	Inclusion and/or Exclusion Criteria Definition of Fall	Sample Characteristics	Gait Assessment	Gait Parameters Related to Falls (Fallers vs. Non-Fallers)	Gait Parameters Not Related to Falls (Fallers vs. Non-Fallers)
Iwata et al., 2014 [62]	Exclusion and inclusion criteria: not reported. Definition of fall: any unintended contact with a supporting surface.	Community-dwelling elderly: 28 fallers (≥1 fall in past year; 9 males; 76.0 ± 5.3 years). 84 non-fallers (19 males; 73.5 ± 6.1 years).	Maximum gait speed was measured using a floor-based photocell gait analysis system over a 5 m course.		Maximum gait speed (m/s): 1.8 vs. 1.9.
Kobayashi et al., 2014 [63]	Inclusion criteria: walk independently; normal or corrected-to-normal vision; and no history of neuromuscular disease. Definition of fall: not reported.	Community-dwelling elderly: 18 fallers (≥1 fall in past year; 67.3 ± 3.1 years). 19 non-fallers (67.1 ± 3.3 years).	Subjects walked at preferred gait speed on a 10 m walkway, during 5 trials. Gait kinematic and ground reaction force data were assessed using an optoelectronic motion analysis system (200 Hz) and six force platforms (1000 Hz). Principal component analysis was used to analyze the relationship between the risk of falling and the joint kinematics of the lower limbs. MTC was analyzed.	Gait speed (m/s): 1.21 vs. 1.33; stance time variability (SD; s): 0.014 vs. 0.009. Fallers exhibited greater variability in the hip, knee, and ankle in all planes during the entire swing phase. Fallers exhibited greater variability in the hip and ankle in the frontal plane during the entire stance phase. Fallers exhibited smaller hip flexion and ankle dorsiflexion angles between the mid-stance and late-stance phases. Fallers exhibited larger ankle inversion between the mid-stance and late-stance phases. Fallers exhibited smaller hip abduction during the mid-stance phase. Variability in the joint kinematics is the key characteristic that affects the risk of falling while walking.	MTC (cm): 4.30 vs. 4.24; step length (m): 0.63 vs. 0.67; step width (m): 0.10 vs. 0.09; stance time (s): 0.60 vs. 0.58; swing time (s): 0.41 vs. 0.41; stance phase (%): 64.8 vs. 63.9; MTC variability (SD; cm): 0.27 vs. 0.29; gait speed variability (SD; m/s): 0.04 vs. 0.03; step length variability (SD; m): 17.8 vs. 13.5; step width variability (SD; cm): 16.3 vs. 16.8; swing time variability (SD; s): 0.02 vs. 0.01; stance phase variability (SD; %): 0.90 vs. 0.76.
König et al., 2014 [64]	Exclusion criteria: body mass index < 18 or >33 kg/m^2; alcoholism; type-1 diabetes; cardiac infarct; chronic hepatitis; celiac and malabsorption diseases; rheumatoid arthritis; cancer; treated for more than 3 months or under treatment with oral corticosteroids; hyperparathyroidism; hyperthyroidism; neurological diseases affecting neuromuscular system; peripheral neurologic diseases; fractures or osteosyntheses; total hip replacement (less than 6 months);	Community-dwelling elderly: 38 fallers (≥1 fall in past year; 69.2 ± 4.8 years). 42 non-fallers (68.9 ± 4.5 years).	Subjects walked barefoot at preferred gait speed on a 10 m walkway during 6 trials. Gait kinematic parameters were assessed using an optoelectronic motion analysis system (200 Hz). Principal component analysis was used.	Temporal variability and mean spatial gait parameters.	

Table 1. *Cont.*

Study	Inclusion and/or Exclusion Criteria Definition of Fall	Sample Characteristics	Gait Assessment	Gait Parameters Related to Falls (Fallers vs. Non-Fallers)	Gait Parameters Not Related to Falls (Fallers vs. Non-Fallers)
	unable to follow instructions or unable to walk 10 m without a walking aid; or participation in another study at the same time. Definition of fall: not reported.				
Mignardot et al., 2014 [65]	Exclusion criteria: refusal or lack of capacity to give consent or hospitalized at the time of screening. Inclusion criteria: 66–75 years; living at home; never fallen; and ability to walk without assistance for at least 30 s. Definition of fall: unintentionally fall on the ground or lower level, not as a result of a major intrinsic event (e.g., as a stroke) or overwhelming hazard.	Community-dwelling elderly: 72 fallers (≥1 fall in 2 years follow-up: 35 males; 71.1 ± 2.7 years). 187 non-fallers (72 males; 69.4 ± 2.5 years).	Subjects walked with their own shoes at preferred gait speed along a 30 m walkway. A tri-axial accelerometer sensor was used (100 Hz). Principal component analysis was used to assess the relationship between gait variables and fall status.	PC1—global kinetics of gait pattern (mechanical power and spatiotemporal variables): fallers (+0 to +6 months) differed from non-fallers and fallers (+6 to +12 months); PC1 had predictive power for the first fall onset during the first six months after the initial screening. PC2—global gait regularity: fallers (+6 to +12 months) differed from non-fallers (+0 to +6 months); PC2 had predictive power for the first fall onset between the 6th and 12th months after initial screening.	PC3—stride time: there was no significant difference between fallers and non-fallers on PC3; PC3 did not have any predictive power for the first fall onset.
Cebolla et al., 2015 [66]	Inclusion criteria: ≥60 years; able to perform activities of daily living and walk independently; and no orthopedic problems (e.g., surgery or fractures) or other health problems that impair physical tests. Definition of fall: unintentionally coming to rest on the ground or other lower level, whether or not it produced an injury.	Community-dwelling elderly (13 males): 20 fallers (≥1 fall in past year; 68.0 ± 6.9 years). 42 non-fallers (65.5 ± 4.1 years).	Subjects walked at preferred gait speed on an 8 m walkway during 10 trials. Gait kinematic parameters were assessed using a 6-camera optoelectronic motion analysis system (100 Hz). MTC was analyzed.	MTC (mm): 40 vs. 43.	Stride length (m): 1.11 vs. 1.17; stride time (s): 1.07 vs. 1.08; cadence (strides/s): 0.93 vs. 0.93; gait speed (m/s): 1.04 vs. 1.08; heel vertical velocity at heel strike (m/s): 0.70 vs. 0.76.
MacAulay et al., 2015 [67]	Exclusion criteria: Geriatric Depression Scale score ≥ 6 or neurologic or untreated health disorders (e.g., cerebrovascular disease, Parkinson's disease, traumatic brain injury). Definition of fall: subject unexpectedly lost his stability and unintentionally came unto rest on the ground or other object; events in which participants were able to regain their stability did not count as a fall.	Community-dwelling elderly (128 males): 81 fallers (≥1 fall in past year; 69.9 ± 6.8 years). 312 non-fallers (70.1 ± 6.6 years).	Subjects walked at preferred gait speed along a 6 m pressure-sensing walkway. Four trials were collected.	Stride length (lower fallers).	Step time (s): 1.08 vs. 1.08.

Table 1. *Cont.*

Study	Inclusion and/or Exclusion Criteria Definition of Fall	Sample Characteristics	Gait Assessment	Gait Parameters Related to Falls (Fallers vs. Non-Fallers)	Gait Parameters Not Related to Falls (Fallers vs. Non-Fallers)
Rispens et al., 2015 [68]	Inclusion criteria: Mini Mental Status Examination score > 18 and able to walk at least 20 m with a walking aid. Definition of fall: event that resulted in unintentionally coming to rest on the ground or other lower level.	Community-dwelling elderly (33 males; 78.4 ± 7.8 years): 41 fallers (≥1 fall in past year). 69 non-fallers.	Subjects wear a portable tri-axial accelerometer sensor (100 Hz) for 2 separate weeks. It was attached with an elastic belt around the waist and set along the lumbar spine. Subjects were instructed to wear the accelerometer at all times, except during water activities. Intra-class correlation was used.	Gait speed; gait speed variability (SD); stride time; stride time variability (SD); gait symmetry (harmonic ratio); and gait smoothness (index of harmonicity) were associated with the number of falls in the past year.	Cadence variability (SD).
Wright et al., 2015 [69]	Inclusion criteria: able to walk at least 100 m without the use of a gait aid and no neurological disease, head trauma, musculoskeletal impairment, or visual impairment not correctable by lenses. Definition of fall: a loss of balance resulting in the body, or part of the body, coming to rest on the ground.	Community-dwelling elderly: 14 "trip" fallers (≥1 trip fall in past year; 4 males; 71 ± 6 years). 10 "slip" fallers (≥1 slip fall in past year; 4 males; 68 ± 5 years). 16 non-fallers (6 males; 72 ± 5 years).	Subjects walked at preferred gait speed along a walkway. Kinematic and ground reaction forces data were collected during 3 trials using a 14-camera optoelectronic motion analysis system (60 Hz) and two force platforms (120 Hz). CoM and CoP data were calculated.	Differences between fallers (both groups) and non-fallers: CoM–CoP at heel strike (cm): 14.3 vs. 15.3 vs. 12.0. Differences between "slip" fallers and non-fallers: CoM–CoP at foot flat (cm): −14.9 vs. −10.3. Differences between "trip" fallers and "slip" fallers: CoM–CoP at mid-swing (cm): 0.9 vs. 1.2.	"Trip" fallers vs. "slip" fallers vs. non-fallers: gait speed (m/s): 1.19 vs. 1.22 vs. 1.14; stride time (s): 1.06 vs. 1.10 vs. 1.10; stride length (m): 1.26 vs. 1.34 vs. 1.26; CoM–CoP at toe-off (cm): −15.1 vs. −16.5 vs. −14.3; CoM–CoP at late swing (cm): 13.4 vs. 13.2 vs. 11.0; peak braking force (% body mass): −15.9 vs. −16.5 vs. −15.1; instant of peak braking force (% gait cycle): 10.8 vs. 11.2 vs. 10.9; peak propulsive force (% body mass): 17.3 vs. 19.3 vs. 17.1; instant of peak propulsive force (% gait cycle): 54.0 vs. 54.1 vs. 54.3.
Bounyong et al., 2016 [70]	Exclusion and inclusion criteria: not reported. Definition of fall: not reported.	Community-dwelling elderly (8 males; 72.3 ± 6.1 years): 17 fallers (≥1 fall in past year). 35 non-fallers.	Subjects walked 6 trials at preferred gait speed along a 5 m walkway. EMG of rectus femoris, biceps femoris, tibialis anterior, and gastrocnemius were collected (1024 Hz). Co-contraction index was determined based on EMG.	Co-contraction index (between tibialis anterior and gastrocnemius) during stance phase (%): 61.8 vs. 57.5.	
Fujimoto & Chou, 2016 [71]	Inclusion criteria: no history or clinical evidence of neurological, musculoskeletal, or other medical conditions (neurological pathology, head trauma, cerebrovascular accident, vestibular dysfunction, or visual impairment uncorrectable by lenses). Definition of fall: not reported.	Community-dwelling elderly: 15 fallers (≥2 falls in past year; 3 males; 71.9 ± 4.3 years). 15 non-fallers (6 males; 70.0 ± 3.2 years).	Subjects walked barefoot at preferred gait speed along a 10 m walkway. Kinematic data were collected during 6 trials using an 8-camera optoelectronic motion analysis system (60 Hz). CoM was calculated.	CoM position at toe-off (m/s): −0.30 vs. −0.47; CoM mean velocity (m/s): 1.03 vs. 1.29; CoM mean velocity at toe-off (m/s): 1.29 vs. 1.61; CoM acceleration peak prior to toe-off (m/s): 0.38 vs. 0.49; CoM AP acceleration peak (lower fallers).	

Table 1. *Cont.*

Study	Inclusion and/or Exclusion Criteria Definition of Fall	Sample Characteristics	Gait Assessment	Gait Parameters Related to Falls (Fallers vs. Non-Fallers)	Gait Parameters Not Related to Falls (Fallers vs. Non-Fallers)
Ihlen et al., 2016 [72]	Exclusion and inclusion criteria: not reported. Definition of fall: not reported.	Community-dwelling elderly (78.4 ± 4.7 years): 32 fallers (≥2 falls in past year). 39 non-fallers.	Subjects wear a tri-axial accelerometer (100 Hz) for 3 days over the lower back. The refined composite multiscale entropy and refined multiscale permutation entropy were applied to trunk acceleration and velocity signals in the AP, ML, and vertical directions.	Refined composite multiscale entropy is higher for non-fallers compared to fallers for trunk AP, ML, and vertical acceleration. Refined multiscale permutation entropy is higher for non-fallers compared to fallers for trunk ML acceleration in the intermediate and large scales. Refined multiscale permutation entropy is lower for non-fallers compared to fallers for trunk vertical acceleration in the intermediate and large scales.	
Howcroft et al., 2016 [73]	Exclusion criteria: cognitive disorder (self-reported) or unable to walk for 6 min without an assistive device. Inclusion criteria: ≥65 years. Definition of fall: event that results in a subject coming to rest unintentionally on the ground or other lower level, excluding falls resulting from a stroke or overwhelming hazard.	Community-dwelling elderly: 24 fallers (≥1 fall in past 6 months; 13 males; 76.3 ± 7.0 years). 76 non-fallers (31 males; 75.3 ± 6.6 years).	Subjects walked 7.62 m while wearing pressure-sensing insoles (120 Hz) and tri-axial accelerometers on the head, pelvis, and left and right shanks (50 Hz). CoP data were analyzed. Maximum Lyapunov exponent, ratio of even to odd harmonics, SD, and CV are used to analyze data variability.	Head variability (SD; higher fallers); ratio of even to odd harmonics pelvis AP (lower fallers).	CoP ML deviation time (s): 0.03 vs. 0.03; minimum CoP velocity (m/s): 0.03 vs. 0.03; mean CoP velocity (m/s): 0.30 vs. 0.30; median CoP velocity (m/s): 0.24 vs. 0.21; gait speed (m/s): 1.24 vs. 1.20; cadence (steps/min): 112 vs. 111; stride time (s): 1.09 vs. 1.09; stance time (s): 0.71 vs. 0.72; swing time (s): 0.38 vs. 0.38; stride time (CV; %): 3 vs. 3; stance time (CV; %): 5 vs. 6; swing time (CV; %): 8 vs. 11; stance phase (%): 64.6 vs. 65.9; double-support phase (%): 14.6 vs. 15.9; CoP AP displacement (CV; %): 495 vs. 463; CoP ML displacement (CV; %): 650 vs. 666; Impulse during foot-strike to first peak (Ns/kg): 1.20 vs. 1.20; Impulse during MTC to second peak (Ns/kg): 1.47 vs. 1.67; Impulse during second peak to foot-off (Ns/kg): 0.97 vs. 1.08; Impulse during foot-strike to MTC (Ns/kg): 2.35 vs. 2.40; Impulse during MTC to foot-off (Ns/kg): 2.36 vs. 2.67; Impulse during foot-strike to foot-off (Ns/kg): 4.65 vs. 4.99; maximum Lyapunov exponent.

Table 1. *Cont.*

Study	Inclusion and/or Exclusion Criteria Definition of Fall	Sample Characteristics	Gait Assessment	Gait Parameters Related to Falls (Fallers vs. Non-Fallers)	Gait Parameters Not Related to Falls (Fallers vs. Non-Fallers)
Rinaldi & Moraes, 2016 [74] Rinaldi et al., 2017 [75]	Exclusion criteria: Mini Mental Status Examination score < 24; vestibular dysfunction; or unable to walk without assistance. Inclusion criteria: no history of neurological or musculoskeletal disorders and no incorrigible visual impairment. Definition of fall: event in which a subject comes unintentionally to the ground or to some lower level.	Community-dwelling elderly women: 15 fallers (≥1 fall in past year; 70.1 ± 5.1 years). 15 non-fallers (71.8 ± 5.8 years).	Subjects walked at preferred gait speed. Gait kinematic data were assessed using an 8-camera optoelectronic motion analysis system (100 Hz). CoM data were analyzed.	Gait speed (m/s): 1.06 vs. 1.23; step width (m): 0.09 vs. 0.06; step time (s): 0.62 vs. 0.52; gait speed (m/s): 0.64 vs. 0.93; CoM AP velocity (m/s): 0.39 vs. 0.75; percentage of CoM AP velocity (%): 60 vs. 30; margin of dynamic stability in AP direction (m): 0.07 vs. 0.02; margin of dynamic stability in ML direction (m): 0.04 vs. 0.01.	
Bizovska et al., 2017 [76]	Exclusion criteria: musculoskeletal problems or injuries and surgical interventions that were performed within 2 years of the study. Inclusion criteria: ≥60 years and ability to stand and walk without any support. Definition of fall: unexpected event in which the participants come to rest on the ground or lower level. Falls related to sports, such as skiing and cycling, and those caused by a great external force were excluded.	Community-dwelling elderly: 38 fallers (≥1 fall in 6 months follow-up; median 70.9 years). 101 non-fallers (median 70.6 years).	Subjects walked at preferred gait speed along a 30 m walkway for 5 min wearing comfortable sports shoes. Tri-axial accelerometers were attached to L5 and shanks (296.3 Hz). The index of complexity, the computed from multiscale entropy, and the Shannon entropy were used to analyze data variability.	ShE: trunk AP direction: 0.34 vs. 0.31; ShE shanks ML direction: 0.41 vs. 0.37.	Gait speed (m/s): 1.22 vs. 1.23; stride time (s): 1.03 vs. 1.05. Shannon entropy: trunk vertical direction: 0.44 vs. 0.43; trunk ML direction: 0.16 vs. 0.17; shanks vertical direction: 0.59 vs. 0.57; shanks AP direction: 0.58 vs. 0.58. Index of complexity: trunk vertical direction: 12.5 vs. 12.4; trunk ML direction: 17.3 vs. 18.0; trunk AP direction: 9.9 vs. 10.2; shanks vertical direction: 9.0 vs. 8.6; shanks ML direction: 15.20 vs. 15.20; shanks AP direction: 8.5 vs. 8.5. Computed from multiscale entropy.
de Melker Worms et al., 2017 [77]	Exclusion criteria: Mini Mental Status Examination score < 25; rheumatoid arthritis in lower extremities; cerebral vascular disease; Parkinson's disease; peripheral neuropathy; cardiac arrest; bypass treatment; any other neurological or cardiovascular impairment; or unable to walk for 10 min without a walking aid. Definition of fall: event in which a subject unintentionally comes to rest on the ground or other lower level.	Community-dwelling elderly (8 males): 9 fallers (≥1 fall in past year; 70.4 ± 3.6 years). 19 non-fallers (69.3 ± 3.6 years).	Subjects walked at 1 m/s on a treadmill. Two bouts of 5 min of walking. Gait kinematic data were assessed using a 10-camera optoelectronic motion analysis system. CoM was analyzed.	Stance time (CV; %): 3.5 vs. 3.0; local divergence exponent of the CoM velocity: 0.97 vs. 0.88.	Step length (m): 0.51 vs. 0.55; step width (m): 0.15 vs. 0.13; stance time (s): 0.69 vs. 0.73; swing time (s): 0.38 vs. 0.41; step length (CV; %): 4.5 vs. 4.2; step width (CV; %): 15.6 vs. 18.6; swing time (CV; %): 4.9 vs. 4.4.

Table 1. *Cont.*

Study	Inclusion and/or Exclusion Criteria Definition of Fall	Sample Characteristics	Gait Assessment	Gait Parameters Related to Falls (Fallers vs. Non-Fallers)	Gait Parameters Not Related to Falls (Fallers vs. Non-Fallers)
de Melker Worms et al., 2017 [78]	Exclusion criteria: Mini Mental State Examination score < 25. Inclusion criteria: ≥65 years and able to walk independently for 10 min. Definition of fall: event in which a subject unintentionally comes to rest on the ground or other lower level.	Community-dwelling elderly: 8 fallers (≥1 fall in past year). 17 non-fallers.	Subjects walked at 1 m/s on a treadmill. Two bouts of 5 min of walking and two slips were induced. Gait kinematic data were assessed using a 10-camera optoelectronic motion analysis system (100 Hz).		Step length of the recovery step; step width of the recovery step; step length variability in the recovery step (CV); step width variability in the recovery step (CV).
Júnior et al., 2017 [80]	Exclusion criteria: unable to walk without help; severe impairment of stability; or Mini Mental Status Examination score < 13 for elderly illiterate, <18 for 1–7 years of education, <26 for ≥8 years of education. Definition of fall: event in which a subject comes to rest on the ground or lower level.	Community-dwelling elderly: 27 fallers (1 fall in past 6 months; 1 male; 68.0 ± 5.7 years). 35 non-fallers (11 males; 68.0 ± 4.8 years).	Subjects walked at preferred gait speed during 3 trials along an 8 m pressure-sensing walkway.		Gait speed (m/s): 1.12 vs. 1.27 (statistical tendency for difference, $p = 0.060$); cadence (steps/min): 113 vs. 112; step length (m): 0.60 vs. 0.63; stride time (s): 1.06 vs. 1.07; single support phase (%): 37.6 vs. 38.4; stride time variability (CV; %): 2.8 vs. 2.7.
Marques et al., 2017 [79]	Exclusion criteria: musculoskeletal pain, fractures, or severe soft tissue injury during the previous 6 months or neurological, cardiovascular, or respiratory diseases. Definition of fall: any stability disturbance that caused a subject's body to have significant contact with the floor.	Community-dwelling elderly women: 16 fallers (≥1 injury fall in past year; 69.6 ± 8.1 years). 19 non-fallers (66.1 ± 6.2 years).	Subjects walked on a treadmill at preferred gait speed. Kinematic data were collected using a telemetry data acquisition system and gait phases using pressure sensors (2000 Hz). SDNN: SD of all time intervals. SDANN: SD of means of intervals taken every five strides. SDNNi: mean of SD of intervals. rMSSD: root-mean-square of differences between intervals. Triangular index: geometric method calculated based on a histogram of intervals.	Stance time: SDNN (higher fallers); SDNNi (higher fallers); rMSSD (higher fallers); CV (higher fallers). Swing time: SDANN (higher fallers). Step time: SDNN (higher fallers); SDNNi (higher fallers); rMSSD (higher fallers); triangular index (higher fallers).	Preferred gait speed (m/s): 0.90 vs. 0.90. Stance time: SDANN; triangular index. Swing time: SDNN; DNNi; rMSSD; CV; triangular index. Step time: SDANN; CV.
Svoboda et al., 2017 [81]	Exclusion criteria: neurological or vestibular diseases or surgery in lower limbs or spine in the last 2 years. Inclusion criteria: ≥60 years; ability to walk without an assistive device; and ability to stand unassisted without any support during common everyday activities. Definition of fall: unexpected event in which the subject comes to rest on the ground or lower level.	Community-dwelling elderly: 31 fallers (≥1 fall in past 6 months; 4 males; 70.9 ± 6.2 years). 94 non-fallers (19 males; 70.4 ± 6.6 years).	Subjects walked barefoot on a 10 m walkway. Each subject participated in 5 trials at preferred, defined (between 1.00 and 1.22 m/s), and fast gait speed. Ground reaction force data were collected using two force platforms. CV and SD were used to analyze variability. CoP ML and AP displacements were calculated.	Preferred gait speed: gait speed (CV; %): 5.9 vs. 5.0. Defined gait speed: gait speed (CV; %): 6.1 vs. 5.0; CoP ML displacement variability during pre-swing (SD; mm): 1.14 vs. 0.85. Fast gait speed: step width variability (CV; %): 27.7 vs. 22.7.	Preferred gait speed: step length: 0.58 vs. 0.59; step width (cm): 9.5 vs. 10.3; step time (s): 0.53 vs. 0.53; gait speed (m/s): 1.11 vs. 1.13; step length (CV; %): 3.1 vs. 3.1; step width (CV; %): 23.7 vs. 24.3; step time (CV; %): 4.1 vs. 3.5; CoP ML and AP displacements variabilities during loading response (mm): 3.11 vs. 3.20; 5.36 vs. 5.03; CoP ML and AP

Table 1. *Cont.*

Study	Inclusion and/or Exclusion Criteria Definition of Fall	Sample Characteristics	Gait Assessment	Gait Parameters Related to Falls (Fallers vs. Non-Fallers)	Gait Parameters Not Related to Falls (Fallers vs. Non-Fallers)
					displacements variabilities during mid-stance (mm): 0.16 vs. 0.16; 0.45 vs. 0.47; CoP ML and AP displacements variabilities during terminal stance (mm): 0.15 vs. 0.15; 0.56 vs. 0.58; CoP ML and AP displacements variabilities during pre-swing (mm): 0.99 vs. 0.87; 3.16 vs. 2.37. Defined gait speed: step length: 0.58 vs. 0.59; step width (cm): 9.7 vs. 10.0; step time (s): 0.52 vs. 0.53; gait speed (m/s): 1.12 vs. 1.11; step length (CV; %): 3.3 vs. 3.0; step width (CV; %): 24.6 vs. 26.3; step time (CV; %): 4.1 vs. 3.7; COP ML and AP displacements variabilities during loading response (mm): 3.11 vs. 3.10; 6.03 vs. 5.18; CoP ML and AP displacements variabilities during mid-stance (mm): 0.15 vs. 0.16; 0.43 vs. 0.44; CoP ML and AP displacements variabilities during terminal stance (mm): 0.15 vs. 0.15; 0.57 vs. 0.57; CoP AP displacement variability during pre-swing (mm): 3.59 vs. 2.22. Fast gait speed: step length: 0.65 vs. 0.66; step width (cm): 9.6 vs. 10.4; step time (s): 0.43 vs. 0.44; gait speed (m/s): 1.53 vs. 1.50; step length (CV; %): 3.7 vs. 3.3; step time (CV; %): 3.8 vs. 3.6; gait speed (CV; %): 5.2 vs. 4.6; CoP ML and AP displacements variabilities during loading response (mm): 4.44 vs. 3.91; 8.54 vs. 8.06; CoP ML and AP displacements variabilities during mid-stance (mm): 0.25 vs. 0.23; 1.06 vs. 0.97; CoP ML and AP displacements variabilities during terminal stance (mm): 0.17 vs. 0.17; 0.81 vs. 0.69.

Table 1. *Cont.*

Study	Inclusion and/or Exclusion Criteria Definition of Fall	Sample Characteristics	Gait Assessment	Gait Parameters Related to Falls (Fallers vs. Non-Fallers)	Gait Parameters Not Related to Falls (Fallers vs. Non-Fallers)
Allen & Franz, 2018 [82]	Exclusion criteria: body mass index $\geq$ 30 kg/m^2; sedentary lifestyle; neurologic or orthopedic diseases; taking medication that causes dizziness; or normal or corrected to normal vision. Definition of fall: unintentionally coming to the ground or some lower level, and other than as a sustaining violent blow, loss of consciousness, or sudden onset of paralysis.	Community-dwelling elderly: 10 fallers ($\geq$1 fall in past year; 3 males; 77.7 $\pm$ 7.7 years). 11 non-fallers (5 males; 75.1 $\pm$ 5.8 years).	Preferred gait speed on a walkway was calculated using two photocells. Subjects walked at preferred gait speed on a treadmill. Kinematic data were assessed using a 14-camera optoelectronic motion analysis system (100 Hz). EMG of leg was recorded (1000 Hz).	Muscle synergy: 2.7 vs. 3.1. Variance in leg muscle recruitment accounted for by one module (larger fallers).	Preferred gait speed (m/s): 1.04 vs. 1.17 (statistical tendency for the difference, $p = 0.060$). Hip flexion and adduction angular position peaks; knee flexion and ankle dorsiflexion angular position peaks.
Benson et al., 2018 [83]	Exclusion criteria: able to walk without an assistive device for 5 min or Mini Mental State Examination score < 22. Definition of fall: unintentionally coming to rest on the ground.	Community-dwelling elderly: 10 fallers ($\geq$1 fall in past 6 months; 5 males; 75.3 years). 10 non-fallers (3 males; 71.9 years).	Subjects walked at preferred gait speed on a treadmill with laboratory shoes. Kinematic data were assessed using a 10-camera optoelectronic motion analysis system (100 Hz).		Knee displacement; knee angle peak.
Howcroft et al., 2018 [86]	Exclusion criteria: cognitive disorder (self-reported) or unable to walk for 6 min without an assistive device. Inclusion criteria: $\geq$65 years and without a fall in the 6 months before evaluation. Definition of fall: event that results in a subject coming to rest unintentionally on the ground or other lower level, excluding falls resulting from a stroke or overwhelming hazard.	Community-dwelling elderly: 28 fallers ($\geq$1 fall in 6 months follow-up; 14 males; 75.0 $\pm$ 8.2 years). 47 non-fallers (17 males; 75.3 $\pm$ 5.5 years).	Subjects walked 7.62 m while wearing pressure-sensing insoles (120 Hz) and tri-axial accelerometers on the head, pelvis, and left and right shanks (50 Hz). CoP AP and ML displacements were calculated. Fast Fourier transform first quartile and CV were used to analyze data variability.	Fast Fourier transform first quartile of left shank ML displacement was lower in fallers. Fast Fourier transform first quartile of right shank vertical displacement was lower in fallers.	Gait speed (m/s): 1.17 vs. 1.22; cadence (steps/min): 110 vs. 112; stride time (s): 1.11 vs. 1.09; stance time (s): 0.73 vs. 0.72; swing time (s): 0.38 vs. 0.37; stride time (CV; %): 3.0 vs. 3.0; stride time symmetry index: 2.13 vs. 2.18; lateral deviation length (mm): 0.9 vs. 1.0; ML deviation time (s): 0.03 vs. 0.03; CoP minimum velocity (m/s): 0.03 vs. 0.03; CoP mean velocity (m/s): 0.28 vs. 0.29; CoP median velocity (m/s): 0.25 vs. 0.25; CoP AP (CV; %): 4.9 vs. 4.5; CoP ML (CV; %): 6.6 vs. 6.7. Impulse foot-strike to 1st peak (Ns/kg): 1.22 vs. 1.20; Impulse 1st peak to minimum (Ns/kg): 1.22 vs. 1.27; impulse minimum to 2nd peak (Ns/kg): 1.83 vs. 1.58; impulse 2nd peak to foot-off (Ns/kg): 1.14 vs. 1.05; impulse foot-strike to minimum (Ns/kg): 2.36 vs. 2.44; impulse minimum to foot-off (Ns/kg): 2.89 vs. 2.56; impulse foot-strike to foot-off (Ns/kg): 5.19 vs. 4.89.

Table 1. *Cont.*

Study	Inclusion and/or Exclusion Criteria Definition of Fall	Sample Characteristics	Gait Assessment	Gait Parameters Related to Falls (Fallers vs. Non-Fallers)	Gait Parameters Not Related to Falls (Fallers vs. Non-Fallers)
Kwon et al., 2018 [84]	Exclusion criteria: not walking independently without assistance devices or any diseases that affected physical activity (e.g., musculoskeletal disease, neurological disease, and cardiovascular disorders). Definition of fall: not reported.	Community-dwelling elderly: 38 fallers (≥1 fall in past year; 10 males; 74.8 ± 5.7 years). 38 non-fallers (74.5 ± 5.0 years).	Subjects walked at preferred gait speed along a pressure-sensing walkway. Three trials were recorded for each subject. SD was calculated.	Gait speed (m/s): 1.01 vs. 1.09; swing phase (%): 36.6 vs. 37.7; stance phase (%): 63.4 vs. 62.4; double support (%): 26.6 vs. 24.5; step time variability (SD; s): 0.04 vs. 0.02; step length (m): 0.54 vs. 0.57; time peak force at maximal weight acceptance and mid-stance (s): 0.22 vs. 0.19; 0.37 vs. 0.34.	Single support (%): 36.9 vs. 37.5; step time (s): 0.55 vs. 0.53; step length variability (SD; cm): 2.6 vs. 2.7; foot progression angle (°): 8.0 vs. 6.5; peak force at maximal weight acceptance, mid-stance, and push-off (N/kg): 1.07 vs. 1.07; 0.82 vs. 0.81; 1.02 vs. 1.01; time to reach push-off (s): 0.56 vs. 0.55.
Marques et al., 2018 [85]	Inclusion criteria: 60–80 years; no use of any gait assistive device; no history of progressive or non-progressive neurological disease; normal cognition; normal or corrected vision; and no cardiovascular, metabolic, or musculoskeletal dysfunction that could impact the safe performance of data collection. Definition of fall: not reported.	Community-dwelling elderly: 53 fallers (≥1 fall in past year; 67.6 ± 5.3 years). 49 non-fallers (64.5 ± 7.1 years).	Subjects walked at preferred gait speed along a 14 m walkway. Six to ten trials were recorded until the subjects completed 50 consecutive strides. Two footswitches were attached on the heel and on the base of the first metatarsus (1000 Hz).	Gait speed (m/s): 1.01 vs. 1.12; stance time (s): 0.58 vs. 0.48; swing time (s): 0.48 vs. 0.57; stride time (s): 1.11 vs. 1.02; double-support time (s): 0.15 vs. 0.10; stride length (m): 1.02 vs. 1.16; stance time variability (SD; s): 0.12 vs. 0.05.	Swing time variability (SD; s): 0.25 vs. 0.17; stride time variability (SD; s): 0.23 vs. 0.21.
Thompson et al., 2018 [88] Qiao et al., 2018 [87]	Exclusion criteria: body mass index ≥ 30 kg/m^2; sedentary lifestyle; orthopedic or neurological condition; or taking medication that causes dizziness. Definition of fall: unintentionally coming to the ground or some lower level for reasons other than a violent blow, loss of consciousness, or sudden onset of paralysis.	Community-dwelling elderly: 11 fallers (≥1 fall in past year; 4 males; 78.3 ± 7.6 years). 11 non-fallers (5 males; 75.3 ± 5.4 years).	Subjects walked at preferred gait speed on a treadmill (in an immersive virtual environment). Kinematic data were assessed using a 14-camera optoelectronic motion analysis system (100 Hz). EMG of leg was recorded (1000 Hz).		Step length (m): 0.62 vs. 0.64; step length variability (SD; cm): 2.7 vs. 2.7; step width (m): 11.2 vs. 12.9; step width variability (SD; cm): 3.3 vs. 2.8. Lower leg antagonist co-activation.
Watanabe, 2018 [89]	Exclusion criteria: women. Inclusion criteria: no history of any musculoskeletal or neurological disorder. Definition of fall: not reported.	Community-dwelling elderly males: 6 fallers (1 fall in past year; 69.0 ± 3.7 years). 7 non-fallers (73.3 ± 6.5 years).	Subjects walked on a treadmill at preferred gait speed for 20 min. Lower extremity kinematics were collected using a 6-camera optoelectronic motion analysis system (100 Hz). EMG data (rectus femoris muscle) were recorded. MTC was analyzed.	MTC decreased with time in non-fallers but not in fallers.	Preferred gait speed (m/s): 1.28 vs. 1.33; 5–10 min: cadence (steps/min): 118 vs. 120; toe off (% gait cycle): 63.2 vs. 61.1; MTC (% gait cycle): 84.3 vs. 82.4. 15–20 min: cadence (steps/min): 116 vs. 117; toe off (% gait cycle): 63.3 vs. 61.2; MTC (% gait cycle): 84.0 vs. 82.3. Variability in the rectus femoris activation decreased with time in fallers and non-fallers.

Table 1. *Cont.*

Study	Inclusion and/or Exclusion Criteria Definition of Fall	Sample Characteristics	Gait Assessment	Gait Parameters Related to Falls (Fallers vs. Non-Fallers)	Gait Parameters Not Related to Falls (Fallers vs. Non-Fallers)
Bueno et al., 2019 [90]	Inclusion criteria: woman; ≥65 years; independent walking without aids; absence of previous surgeries in lower limbs, pelvis, or spine; Mini Mental Status Examination score > 14; body mass index ≥ 30 kg/m^2; rheumatoid arthritis; neuromuscular or neurodegenerative diseases; diabetes mellitus; and no visual impairment. Definition of fall: unexpected event in which the subject comes to rest on the ground or lower level (excluded coming to rest against furniture or wall).	Community-dwelling elderly women: 10 recurrent fallers (≥2 falls in past year; 71.0 ± 6.8 years). 12 fallers (1 fall in past year; 72.7 ± 5.6 years). 27 non-fallers (72.5 ± 6.8 years).	Subjects walked barefoot at preferred gait speed along a 9 m walkway. Kinematic data were collected using a 7-camera optoelectronic motion analysis system (120 Hz).		Fallers vs. Recurrent fallers vs. Non-fallers: gait speed (m/s): 1.02 vs. 0.99 vs. 1.00; cadence (step/min): 110 vs. 113 vs. 109; stride length (m): 1.10 vs. 1.04 vs. 1.02; hip flexion/extension (°): 8.2 vs. 7.1 vs. 8.2; knee flexion/extension (°): 7.1 vs. 8.1 vs. 7.6; ankle dorsiflexion/plantarflexion (°): 4.7 vs. 4.5 vs. 4.5; hip adduction/abduction (°): 4.7 vs. 5.2 vs. 4.2; hip rotation (°): 6.4 vs. 6.9 vs. 6.6; foot progression angle (°): 5.8 vs. 6.1 vs. 6.0.
Mak et al., 2019 [92]	Exclusion criteria: Mini Mental Status Examination score < 24; neurological impairment; acquired static visual acuity worse than 20/40; or unable to walk independently indoors. Definition of fall: not reported.	Community-dwelling elderly (40 males; 70 3 ± 4.8 years): 37 fallers (≥1 fall in past year; 7 males; 70.7 ± 5.0 years). 97 non-fallers (33 males; 70.1 ± 4.7 years).	Subjects walked along a 6 m walkway at preferred gait speed. Average muscle activity indices of lower limb co-contractions were measured using surface EMG.	Shank and thigh muscle co-contractions were higher in fallers. Lower limb muscle co-contractions were higher in fallers.	
Gillain et al., 2019 [91]	Exclusion criteria: ≥1 fall in the past year; use of a walking aid; gait disorders or an increased fall risk related to neurological or osteoarticular diseases; dementia; hip or knee prosthesis in the previous year; pain when walking; acute respiratory or cardiac illness; recent hospitalization; untreated or uncontrolled comorbidities; use of neuroleptic and sedative drugs; or presence of a cardiac pacing device. Definition of fall: unexpected event in which the subject comes to rest on the ground.	Community-dwelling elderly (48 males; 71.3 ± 5.4 years): 35 fallers (≥1 fall in 2 years follow-up; 18 males; 72.0 ± 6.9 years). 61 non-fallers (30 males; 70.9 ± 4 3 years).	Subjects walked with their own shoes at preferred and fast gait speed. Kinematic data were collected using a tri-axial accelerometer attached to the lumbar position and a 24-camera optoelectronic motion analysis system (120 Hz). Variability was analyzed using the CV. MTC was analyzed.	Preferred gait speed: stride length (m): 1.30 vs. 1.37. Fast gait speed: gait speed (m/s): 1.64 vs. 1.74; stride length (m): 1.47 vs. 1.60;	Preferred gait speed: gait speed (m/s): 1.24 vs. 1.31. Preferred gait speed and fast gait speed: cadence (stride/s): 0.96 vs. 0.96; 1.10 vs. 1.08; MTC (mm): 17.3 vs. 17.8; 18.0 vs. 20.6; median MTC (mm): 17.4 vs. 17.7; 18.7 vs. 20.8; MTC variability (SD; mm): 5.0 vs. 4.1; 4.5 vs. 4.5; MTC variability (CV; %): 27.0 vs. 24.3; 26.1 vs. 27.6; MTC-minimum (mm): 10.8 vs. 12.0; 9.8 vs. 9.1.

Table 1. *Cont.*

Study	Inclusion and/or Exclusion Criteria Definition of Fall	Sample Characteristics	Gait Assessment	Gait Parameters Related to Falls (Fallers vs. Non-Fallers)	Gait Parameters Not Related to Falls (Fallers vs. Non-Fallers)
Yamagata et al., 2019 [93]	Exclusion criteria: neurological disorders or musculoskeletal injuries that would affect performance or inability to walk without assistance. Definition of fall: an unexpected event during which the subjects come to rest on the ground or other lower level.	Community-dwelling elderly: 12 fallers (≥1 fall in 1 year follow-up; 78.0 ± 4.7 years). 16 non-fallers (73.8 ± 7.9 years).	Subjects walked at preferred gait speed on a 6 m walkway. Kinematic data were assessed by an 8-camera optoelectronic motion analysis system (100 Hz). CoM was calculated. SD analyzed variability.	Gait speed (m/s): 1.1 vs. 1.3; CoM distance at toe-off and at heel strike (cm): 22.9 vs. 25.3; 19.7 vs. 2.26; variability in right shank angle in the sagittal plane during mid-swing (SD; rad): 6.41 vs. 5.21.	Segment angles (foot, shank, thigh, pelvis); segment angles variability (foot, shank, thigh, pelvis). Step width during early swing, mid-swing, and late-swing (m): 0.13 vs. 0.14; 0.15 vs. 0.17; 0.14 vs. 0.16; step width variability during early swing, mid-swing, and late-swing (m): 0.018 vs. 0.023; 0.011 vs. 0.0127; 0.019 vs. 0.020.
Yamagata et al., 2019 [94]	Exclusion criteria: neurological disorders or musculoskeletal injuries that would affect performance or inability to walk without assistance. Definition of fall: an unexpected event during which the subjects come to rest on the ground or other lower level.	Community-dwelling elderly: 10 fallers (≥1 fall in past year; 78.0 ± 2.7 years). 14 non-fallers (75.1 ± 5.4 years).	Subjects walked at preferred gait speed on a 6 m walkway. Kinematic data were assessed by an 8-camera optoelectronic motion analysis system (100 Hz). SD is used to analyze variability. Uncontrolled manifold analysis was used to assess how variability in segmental configurations affects the frontal trajectory of the swing foot.	Step length (m): 0.52 vs. 0.56; right shank angle in early swing (lower fallers); shank angle variability during early and mid-swing (higher fallers); variability of the vertical distance between feet (higher fallers). Fallers: higher variability in segmental configurations in all phases; in ML direction, kinematic synergy was higher during the early and late swing; higher kinematic synergy in the vertical direction.	Gait speed (m/s) (statistical tendency for the difference, $p = 0.060$): 1.01 vs. 1.25; step width (m): 0.11 vs. 0.10; cadence (steps/min): 97 vs. 99; swing phase (%): 36.7 vs. 35.2; stride length ratio: 0.9 vs. 0.9. In all planes, there were no differences in segment angles (foot, shank, thigh, pelvis); segment angle variability (foot, shank, thigh, pelvis). Vertical and ML distances between feet. Variability ML distance between feet.
Gonzalez et al., 2020 [95]	Inclusion criteria: ability to walk one mile at any pace with minimum rest and body mass index = 18.5–24.9 kg/m^2 or ≥30 kg/m^2. Exclusion criteria: use of an assistive device for walking; artificial joint replacement; history of diabetic peripheral neuropathy, neurological conditions that interfere with gait; body mass index = 25–29.9 kg/m^2; compromised range of motion in the lower limb or trunk; untreated hypertension or cardiovascular diseases; or T-score ≤ 2.5 for the femoral neck. Definition of fall: not reported.	Community-dwelling elderly: 16 fallers (≥1 fall in 1-year follow-up). 16 non-fallers.	Subjects walked on a treadmill for 10 min at preferred gait speed. Kinematic data of the 10th thoracic vertebra were assessed by an 8-camera optoelectronic motion analysis system (100 Hz). Short-term exponents were used to analyze the data variability for each direction (larger positive exponents indicate higher instability).		Short-term exponents of the ML, AP, or vertical displacements.

Table 1. *Cont.*

Study	Inclusion and/or Exclusion Criteria Definition of Fall	Sample Characteristics	Gait Assessment	Gait Parameters Related to Falls (Fallers vs. Non-Fallers)	Gait Parameters Not Related to Falls (Fallers vs. Non-Fallers)
Pol et al., 2021 [97]	Exclusion criteria: <60 years; Short Portable Mental Status Questionnaire score < 7; unable to ambulate for at least 10 m without an assistive device; diabetic foot syndrome; neurological diseases; or lower extremity surgery. Definition of fall: unintentionally coming to the ground or other lower surface, not as a result of a major intrinsic event or an overwhelming hazard.	Community-dwelling elderly: 74 fallers (≥1 fall in past year; 20 males; 71.9 ± 4.9 years). 113 non-fallers (61 males; 69.9 ± 5.5 years).	Three trials were recorded for each subject's dominant limb using a two-step gait initiation protocol at a comfortable walking speed. Foot function was assessed using barefoot plantar pressure analysis (50 Hz). CoP was calculated.	CoP excursion index (%): 14.69 vs. 17.58, total pressure–time integral (% body weight * second/cm^2): 3.75 vs. 3.23; pressure–time integral of medial forefoot (% body weight * second/cm^2): 1.84 vs. 1.39.	Pressure–time integral of medial heel (% body weight * second/cm^2): 1.52 vs. 1.44; pressure–time integral of lateral heel (% body weight * second/cm^2): 1.48 vs. 1.36; pressure–time integral of medial midfoot (% body weight * second/cm^2): 0.99 vs. 0.87; pressure–time integral of lateral midfoot (% body weight * second/cm^2): 1.30 vs. 1.21; pressure–time integral of central forefoot (% body weight * second/cm^2): 1.73 vs. 1.71; pressure–time integral of lateral forefoot (% body weight * second/cm^2): 1.75 vs. 1.79; pressure–time integral of hallux (% body weight * second/cm^2): 2.35 vs. 1.80; pressure–time integral of second toe (% body weight * second/cm^2): 0.90 vs. 0.78; pressure–time integral of lateral toes (% body weight * second/cm^2): 0.84 vs. 0.73; Regional CoP velocity–heel, midfoot, forefoot, and toes (cm/s): 25.6 vs. 28.1; 16.1 vs. 18.8; 19.1 vs. 18.1; 43.8 vs. 48.4.
Sadeghi et al., 2021 [98]	Exclusion criteria: need help for walking; difficulties in understanding instructions; or receiving hospice care. Definition of fall: not reported.	Community-dwelling elderly: 13 fallers (>1 fall in past year; 72.5 ± 7.1 years). 13 non-fallers (73.1 ± 7.1 years).	Subjects walk barefoot at preferred gait speed on a 10 m walkway. Kinematic data from 10 gait cycles were collected by a 10-camera optoelectronic motion analysis system (100 Hz).	Cadence (steps/min): 98 vs. 115; gait speed (m/s): 0.74 vs. 1.04; stride time (s): 1.27 vs. 1.05; stride length (m): 0.90 vs. 1.08; double support time (s): 0.30 vs. 0.26. Ankle-to-knee, knee-to-hip, and ankle-to-hip coordination patterns. Less coordination variability in fallers.	Step width (m): 0.10 vs. 0.10; ankle displacement (°): 25 vs. 23; knee displacement (°): 47 vs. 44; hip displacement (°): 42 vs. 41.
Yamagata et al., 2021 [96]	Exclusion criteria: neurological disorders or musculoskeletal injuries that would affect performance or inability to walk without assistance. Definition of fall: an unexpected event during which the subjects come to rest on the ground or other lower level.	Community-dwelling elderly: 10 fallers (≥1 fall in past year; 78.0 ± 2.7 years). 14 non-fallers (75.1 ± 5.4 years).	Subjects walked at preferred gait speed on a 6 m walkway. Kinematic data were assessed by an 8-camera optoelectronic motion analysis system (100 Hz). CoM was calculated. Variance is used to analyze variability.	CoM vertical direction variability (higher fallers).	CoM displacements. CoM ML direction variability.

Table 1. *Cont.*

Study	Inclusion and/or Exclusion Criteria Definition of Fall	Sample Characteristics	Gait Assessment	Gait Parameters Related to Falls (Fallers vs. Non-Fallers)	Gait Parameters Not Related to Falls (Fallers vs. Non-Fallers)
Figueiredo et al., 2022 [99]	Inclusion criteria: ≥80 years; any gender; ability to walk independently; and ability to understand the verbal commands to carry out assessment. Exclusion criteria: uncertain about their history of falls; had been hospitalized for >7 days in the 3 months before assessment; or had a severe orthopedic, neurological, respiratory, cardiovascular, visual, or hearing disease. Definition of fall: unexpected and unexplained event in which the subject inadvertently comes to the ground.	Community-dwelling elderly: 32 fallers (≥1 fall in past 6 months; 7 males; 89.9 ± 4.4 years). 32 non-fallers (5 males; 88.6 ± 4.1 years).	Subjects walked during the Timed Up and Go test. Kinematic data were collected using a tri-axial accelerometer attached between the L5 and S1 vertebrae. Spectral arc length metrics are used to quantify gait smoothness.		Spectral arc length in AP, ML, and vertical directions.
Nascimento et al., 2022 [100]	Inclusion criteria: residing in the community; 60–79 years; and able to walk independently. Exclusion criteria: medical contraindications for submaximal exercise, according to American College of Sports Medicine guidelines, or inability to understand or follow investigation protocols. Definition of fall: not reported.	Community-dwelling elderly: 225 fallers (>1 fall in past year; 70.1 ± 5.6 years). 394 non-fallers (69.2 ± 5.4 years).	Subjects walked a distance of 30 feet at their preferred gait speed. Gait speed is calculated by dividing the distance walked by the time needed. Cadence is calculated by dividing the number of steps taken in space during the 30-foot walk test by the time taken to cover that distance.	Gait speed (m/s): 1.20 vs. 1.28.	Cadence (steps/s): 1.90 vs. 1.92.
Yoshida et al., 2022 [101]	Exclusion criteria: diagnosis of dementia, recent major illness, neurological, sensory, or mobility impairment, or musculoskeletal disorders or injuries. Definition of fall: event that resulted in a person coming to rest unintentionally on the ground or other lower level, not as the result of a major intrinsic event or an overwhelming hazard.	Community-dwelling elderly: 28 fallers (≥1 fall in past year; 7 males; 77.5 ± 4.9 years). 28 non-fallers (12 males; 78.1 ± 5.3 years).	Gait initiation was assessed using two force platforms (1024 Hz). CoP was calculated.	First step length (m): 0.61 vs. 0.66; CoP ML displacement during weight transfer (m): 0.13 vs. 0.12.	Weight transfer time (s): 0.99 vs. 0.96; forward progression time (s): 0.48 vs. 0.49; first step time (s): 0.55 vs. 0.57; ground contact time (s): 0.69 vs. 0.68; first step width (m): 0.17 vs. 0.17; CoP AP displacement during weight transfer (m): 0.043 vs. 0.048; CoP AP displacement during forward progress (m): 0.181 vs. 0.170; CoP ML displacement during forward progress (m): 0.023 vs. 0.023; CoP AP displacement during first step (m): 0.228 vs. 0.225; CoP ML displacement during first step (m): 0.023 vs. 0.021.

Table 1. *Cont.*

Study	Inclusion and/or Exclusion Criteria Definition of Fall	Sample Characteristics	Gait Assessment	Gait Parameters Related to Falls (Fallers vs. Non-Fallers)	Gait Parameters Not Related to Falls (Fallers vs. Non-Fallers)
Baba et al., 2023 [102]	Inclusion criteria: ≥65 years and ability to walk independently without using aids. Exclusion criteria: stroke diagnosis, Parkinson's disease, rheumatism; or history of hip or knee surgery. Definition of fall: unintentional landing on the ground, floor, or lower level.	Community-dwelling elderly: 16 fallers (≥1 fall in past year; 4 males; 84.6 ± 5.7 years). 34 non-fallers (3 males; 81.7 ± 6.1 years).	Subjects walked barefoot on a walkway at preferred gait speed. Kinematic data were collected using an inertial sensor system.	Gait speed (m/s): 0.83 vs. 0.92; foot angle with ground (°): 13.6 vs. 18.3.	Stride time (s): 1.10 vs. 1.03; stride length (m): 0.87 vs. 0.96; cadence (steps/min): 114 vs. 117; stance phase (%): 65.0 vs. 64.7; single support (%): 35.4 vs. 35.9; double support (%): 21.2 vs. 21.0; maximum ankle plantarflexion: 10.8 vs. 9.0; maximum ankle dorsiflexion: 9.7 vs. 9.4; maximum knee flexion: 40.4 vs. 41.7; maximum knee extension: 0.5 vs. 0.5; maximum hip flexion: 25.2 vs. 27.3; maximum hip extension: 13.5 vs. 13.2.

AP—anteroposterior; ApEn—approximate entropy analysis; CoM—center of mass; CoP—center of pressure; CV—coefficient of variation; EMG—electromyography; ML—mediolateral; MTC—minimum toe clearance; rMSSD—root-mean-square of differences between intervals; SampEn—sample entropy analysis; SD—standard deviation; SDANN—standard deviation of means of intervals taken every five strides; SDNN—standard deviation of all time intervals; SDNNi—mean of standard deviations of intervals. *—it is the symbol of multiplication.

Table 2. Characteristics and data of the studies that compared elderly who fell during induced falls and elderly who did not.

Study	Inclusion and/or Exclusion Criteria Definition of Fall	Sample Characteristics	Gait Assessment	Gait Parameters Related to Falls (Fallers vs. Recoveries)	Gait Parameters Not Related to Falls (Fallers vs. Recoveries)
Pavol et al., 1999 [105]	Exclusion criteria: neurological, musculoskeletal, cardiovascular, pulmonary, cognitive, and other systemic disorders; history of repeated falling; or minimum bone mineral density of the femoral neck of 0.65 g/cm^2.	Community-dwelling elderly: 10 fallers (body's subject being at least 50% supported by the safety harness). 39 recoveries.	Subjects walked on a 7 m walkway at their preferred gait speed using a safety harness. One trip was induced using a mechanical obstacle (5.1 cm from the ground). Kinematic data were collected using a 6-camera optoelectronic motion analysis system (60 Hz).	Gait speed (m/s): 1.31 vs. 1.18; step time (s): 0.50 vs. 0.54; step length (% body mass): 43.1 vs. 39.9.	Step width (cm): 8.9 vs. 9.2; trunk flexion (°): 11 vs. 9; gait phase of trip (% stride length): 59 vs. 59.
Lockhart et al., 2003 [106]	Exclusion criteria: based on an exam conducted by a physician to ensure that they were in generally good physical health and a peripheral neuropathy examination (below 50% of the norm).	Community-dwelling elderly (7 males: 75.5 ± 6.8 years): 7 fallers. 7 non-fallers.	Subjects walked on a 20 m circular track for 10 min using a safety harness. Two slips were induced. Four video cameras and two force plates were used to collect 3D data and the ground reaction forces (60 Hz).	Coefficient of friction after a heel strike (horizontal ground reaction force/vertical ground reaction force) was lower in fallers; horizontal heel contact velocity was higher in fallers.	

Table 2. *Cont.*

Study	Inclusion and/or Exclusion Criteria Definition of Fall	Sample Characteristics	Gait Assessment	Gait Parameters Related to Falls (Fallers vs. Recoveries)	Gait Parameters Not Related to Falls (Fallers vs. Recoveries)
Pijnappels et al., 2005 [103]	Exclusion and inclusion criteria: not reported.	Community-dwelling elderly: 7 fallers (when the vertical force in the ropes exceeded 200 N during trials when one obstacle appeared from the ground unexpectedly to catch the subject's swing limb. 1 male; 67.9 ± 2.6 years). 4 non-fallers (3 males; 66.5 ± 3.3 years).	Subjects walked at preferred gait speed over a platform and were tripped several times by an obstacle that appeared from the floor. A safety harness prevented subjects from falling. Kinematic and ground reaction force data were collected using a 4-camera optoelectronic motion analysis system and a force plate (100 Hz).	Total angular momentum; angular momentum at push-off.	Gait speed; stride length; obstacle contact phase; stance time of the support limb; double support time; swing phase time of the recovery limb; hip horizontal displacement; hip height at end push-off; rate of change of moment generation in ankle, knee, and hip; hip extension moment, knee flexion moment and ankle plantarflexion moment peak in the support limb.
Pavol et al., 2001 [107]	Exclusion criteria: neurological, musculoskeletal, cardiovascular, pulmonary, cognitive, and other systemic disorders; history of repeated falling; or minimum bone mineral density of the femoral neck of 0.65 g/cm^2.	Community-dwelling elderly. The recovery attempts were classified as a lowering or an elevating strategy. Elevating strategy: 1 faller. 11 recoveries.	Subjects walked on a 7 m walkway at their preferred gait speed using a safety harness. One trip was induced using a mechanical obstacle (5.1 cm from the ground). Kinematic collected using a 6-camera optoelectronic motion analysis system (60 Hz). Ground reaction forces were collected by two force plates (1000 Hz).	Hip horizontal velocity at time of trip (% body height/s): 92.5 vs. 68.0; hip horizontal velocity at 0.1 s post-trip (%/s): 86.5 vs. 67.4; time from trip to follow-through toe-off (s): 0.40 vs. 0.45; ankle–hip angular position at time of loading (°): 11.9 vs. 8.9; hip height at heel strike (% body height): 51.1 vs. 54.5; trunk inclination from vertical at heel strike (°): 58.5 vs. 37.3; lumbar flexion at heel strike (°): 45.2 vs. 23.1; minimum hip–ankle distance (% body height): 42.4 vs. 47.4; maximum trunk inclination from vertical (°): 83.5 vs. 49.7; maximum lumbar flexion at heel strike (°): 70.3 vs. 35.3.	Trunk inclination at time of trip (°): 14.3 vs. 8.7; hip vertical velocity 0.1 s post-trip (% body height): −8.1 vs. −9.3; lumbar flexion at time of loading (°): 17.2 vs. 6.7; recovery step length (% body height): 51.8 vs. 49.8; recovery stride length (% body height): 93.2 vs. 89.7; obstacle–ankle distance at heel strike (% body height): 32.6 vs. 32.2; minimum hip–ankle distance (% body height): 31.0 vs. 34.5; maximum ankle ground clearance (% body height): 24.0 vs. 22.1; time from trip-to-ground contact (s): 0.40 vs. 0.45; maximum horizontal ankle velocity (% body height/s): 203 vs. 225; average horizontal ankle velocity (% body height/s): 56 vs. 54; ankle–hip angle at heel strike (°): 0.3 vs. −7.6; maximum hip vertical velocity (% body height/s): 20.7 vs. 29.1.
Pavol et al., 2001 [107]	Exclusion criteria: neurological, musculoskeletal, cardiovascular, pulmonary, cognitive, and other systemic disorders; history of repeated falling; or minimum bone mineral density of the femoral neck of 0.65 g/cm^2.	Community-dwelling elderly. The recovery attempts were classified as a lowering or an elevating strategy. Lowering strategy: 5 fallers during-step fall (body's subject being at least 50% supported by safety harness). 3 fallers after-step fall (body's subject being at least 50% supported by safety harness). 26 recoveries.	Subjects walked on a 7 m walkway at their preferred gait speed using a safety harness. One trip was induced using a mechanical obstacle (5.1 cm from the ground). Kinematic data were collected using a 6-camera optoelectronic motion analysis system (60 Hz). Ground reaction forces were collected by two force plates (1000 Hz).	Fallers during-step fall vs. recoveries: hip horizontal velocity at trip (% body height/s): 91.3 vs. 68.2; hip horizontal velocity at 0.1 s post-trip (% body height/s): 94.5 vs. 72.9; time from trip to support limb loading (s): 0.27 vs. 0.16; ankle–hip angle at time of loading (°): 23.6 vs. 9.8; recovery step length (% body height): 36.9 vs. 49.4;	Fallers during-step fall vs. recoveries: trunk inclination at time of trip (°): 7.5 vs. 9.1; hip vertical velocity at 0.1 s post-trip (body height/s): −11.8 vs. −9.8; time from trip to follow-through toe-off (s): 0.49 vs. 0.50; lumbar flexion at time of loading (°): 6.4 vs. 6.1; minimum hip–ankle distance at recovery step (% body height): 31.8 vs. 33.0; maximum ankle ground clearance (% body height): 23.8 vs. 24.7; time from trip to recovery

Table 2. *Cont.*

Study	Inclusion and/or Exclusion Criteria Definition of Fall	Sample Characteristics	Gait Assessment	Gait Parameters Related to Falls (Fallers vs. Recoveries)	Gait Parameters Not Related to Falls (Fallers vs. Recoveries)
				recovery stride length (% body height): 51.4 vs. 59.9; obstacle–ankle distance at ground contact (% body height): 32.0 vs. 39.6; recovery step time (s): 0.21 vs. 0.27; hip height at ground foot contact (% body height): 47.2 vs. 54.5; ankle–hip angle at heel strike (°): 12.7 vs. −10.1; trunk inclination from vertical at heel strike (°): 48.3 vs. 36.0; maximum lumbar flexion at heel strike (°): 23.1 vs. 35.6. Fallers after-step fall vs. recoveries: trunk inclination at trip (°): 18.8 vs. 9.1; trunk inclination from vertical at ground foot contact (°): 55.2 vs. 36.0; lumbar flexion at ground foot contact (°): 38.7 vs. 23.5; maximum hip vertical velocity (% body height): −0.4 vs. 32.2; minimum hip–ankle distance (% body height): 41.0 vs. 47.3; maximum trunk inclination from vertical at ground foot contact (°): 74.6 vs. 46.6; maximum lumbar flexion at ground foot contact (°): 54.4 vs. 35.6.	foot toe-off (s): 0.28 vs. 0.26; time from trip to ground contact (s): 0.49 vs. 0.52; maximum horizontal ankle velocity (% body height/s): 227 vs. 263; average horizontal ankle velocity (% body height/s): 109 vs. 115; lumbar flexion at ground foot contact (°): 22.1 vs. 23.5; Fallers after-step fall vs. recoveries: hip horizontal velocity at time of trip (% body height/s): 79.4 vs. 68.2; hip horizontal velocity at 0.1 s post-trip (% body height/s): 82.2 vs. 72.9; hip vertical velocity at 0.1 s post-trip (% body height/s): −7.2 vs. −9.8; time from trip to support limb loading (s): 0.14 vs. 0.16; time from trip to follow-through toe-off (s): 0.52 vs. 0.50; ankle–hip angle at time of loading (°): 9.1 vs. 9.8; lumbar flexion at time of loading (°): 15.4 vs. 6.1; recovery step length (% body height): 49.1 vs. 49.4; recovery stride length (% body height): 61.7 vs. 59.9; obstacle–ankle distance at heel strike (% body height): 40.0 vs. 39.6; minimum hip–ankle distance (% body height): 28.8 vs. 33.0; time from trip to recovery foot toe-off (s): 0.24 vs. 0.26; time from trip to heel strike (s): 0.51 vs. 0.52; recovery step duration (s): 0.26 vs. 0.270; maximum horizontal ankle velocity (% body height/s): 264 vs. 263; average horizontal ankle velocity (% body height/s): 117 vs. 115; hip height (% body height): 50.9 vs. 54.5, ankle–hip angle (°): −7.8 vs. −10.1.
Espy & Pai, 2007 [108]	Inclusion criteria: ≥65 years.	Community-dwelling elderly: 19 fallers (subjects who fell during an unexpected induced slip). 15 non-fallers (subjects who recover during an unexpected induced slip).	Subjects walked on an instrumented walkway while wearing a safety harness. After 10 unperturbed trials, an unexpected slip was induced under the right heel. Motion data were used to calculate frontal plane variables.		Step width; CoM lateral position; CoM velocity.

Table 2. *Cont.*

Study	Inclusion and/or Exclusion Criteria Definition of Fall	Sample Characteristics	Gait Assessment	Gait Parameters Related to Falls (Fallers vs. Recoveries)	Gait Parameters Not Related to Falls (Fallers vs. Recoveries)
Bhatt et al., 2011 [104]	Exclusion criteria: Folstein Mini Mental Status Examination score < 25 or classified as osteopenic or osteoporotic.	Community-dwelling elderly (44 males): 59 fallers (the force recorded on the safety harness load cell force exceeded 30% of the body weight; 71.6 ± 4.6 years). 56 recoveries (71.4 ± 5.1 years).	Subjects walked for 9–12 trials on a 7 m walkway using their own athletic shoes and were then exposed to one unannounced slip. Kinematic and ground reaction force data were collected during 5 trials using an 8-camera optoelectronic motion analysis system (120 Hz) and one force platform (600 Hz).	Dynamic gait stability: −0.16 vs. −0.13.	
Yang & Pai, 2014 [109]	Exclusion criteria: any known neurological, musculoskeletal, or other systemic disorder that would have affected their postural control.	Community-dwelling elderly: 98 fallers (the force recorded on the safety harness load cell force exceeded 30% of the body weight; 22 males; 71.8 ± 5.5 years). 89 recoveries (37 males; 71.9 ± 4.8 years).	Subjects walked for 20 trials on a 7 m instrumented walkway at preferred gait speed and were then exposed to one unannounced slip. Kinematic data were collected using an 8-camera optoelectronic motion analysis system (120 Hz) and ground reaction force using four force platforms (600 Hz).	Step width (SD; m): 0.031 vs. 0.027; dynamic stability of CoM against backward falling: −0.18 vs. −0.16.	Step length (SD; m): 0.070 vs. 0.062; step time (SD; s): 0.044 vs. 0.041; margin of stability: 0.039 vs. 0.051; Floquet multiplier: 0.422 vs. 0.432; Lyapunov exponent (short-term): 0.671 vs. 0.737; Lyapunov exponent (long-term): 0.034 vs. 0.026.
Sawers et al., 2016 [110]	Inclusion criteria: participants who experienced a "split" slip with the slipping and trailing feet traveling apart were included.	Community-dwelling elderly: 15 fallers (unable to regain their stability after an unexpected induced slip; 2 males; 71.0 ± 2.0 years). 13 recoveries (able to recover their stability and continue walking after an unexpected induced slip; 8 males; 72.0 ± 5.0 years).	Subjects walked on a 7 m walkway at their preferred gait speed using a safety harness. One unexpected slip was induced. Kinematic collected using an optoelectronic motion analysis system (120 Hz). EMG of TA, MG, VL, and BF were recorded (600 Hz).	Slip distance (m): 0.78 vs. 0.61; EMG onset latencies/slip leg (s): VL (right): 0.239 vs. 0.186; BF (right): 0.170 vs. 0.120. Muscle synergies recruited during slip and non-slip trials: 3.7 vs. 4.7.	Slip time (s): 0.82 vs. 0.94; peak slip velocity (m/s): 2.00 vs. 1.84; dynamic stability: −0.18 vs. −0.16; gait speed (m/s): 0.89 vs. 1.00; shank angle (°): 74 vs. 73. EMG onset latencies/slip leg (s): TA (right): 0.173 vs. 0.151; MG (right): 0.234 vs. 0.232; EMG onset latencies/nonslip leg (s): TA (left): 0.162 vs. 0.157; MG (left): 0.215 vs. 0.198, VL (left): 0.165 vs. 0.154; BF (left): 0.150 vs. 0.155; EMG onset peak magnitude/slip leg: TA (right): 2.30 vs. 2.28; MG (right): 2.32 vs. 2.32; VL (right): 2.09 vs. 2.36; BF (right): 3.45 vs. 3.87; EMG Onset peak magnitude/nonslip leg: TA (left): 2.86 vs. 2.58; MG (left): 1.64 vs. 1.87; VL (left): 4.75 vs. 3.56; BF (left): 3.75 vs. 3.22.

Table 2. *Cont.*

Study	Inclusion and/or Exclusion Criteria Definition of Fall	Sample Characteristics	Gait Assessment	Gait Parameters Related to Falls (Fallers vs. Recoveries)	Gait Parameters Not Related to Falls (Fallers vs. Recoveries)
Sawers & Bhatt, 2018 [111]	Inclusion criteria: participants who experienced a feet-forward slip (with both feet moving forward) were included.	Community-dwelling elderly: 12 fallers (when peak force recorded by the load cell in line with the overhead harness exceeded 30% of the subject's body weight; 2 males; 73.0 ± 4.9 years). 13 recoveries (7 males; 74.0 ± 4.1 years).	Subjects walked on a 7 m walkway at their preferred gait speed using a safety harness. One unexpected slip was induced. Kinematic data were collected using an optoelectronic motion analysis system (120 Hz). Ground reaction forces were collected (600 Hz).	Lower limb joint angle: knee flexion (higher flexion fallers). Number of muscle synergies recruited: 4 vs. 5.	Peak slip velocity (m/s): 2.28 vs. 2.14; slip duration (s): 0.68 vs. 0.70; slip distance (m): 0.74 vs. 0.62; shank angle (°): 75.8 vs. 75.0; step length (m): 0.30 vs. 0.32; dynamic stability: −0.124 vs. −0.155; gait speed (m/s): 1.13 vs. 1.02 (statistical tendency for the difference, $p = 0.093$). Lower limb joint angle: hip flexion/extension; knee extension; ankle dorsiflexion/plantar-flexion.
Bruijn et al., 2022 [112]	Exclusion criteria: orthopedic, neuromuscular, cardiac, or visual problems.	Community-dwelling elderly: 5 fallers. 11 recoveries.	Subjects walked on a 12 m walkway at their preferred gait speed using a safety harness. One unexpected slip was induced. Kinematic data were collected using an optoelectronic motion analysis system (50 Hz). Ground reaction forces were collected (1000 Hz).	Sagittal plane forward body rotation at touchdown.	Gait speed (m/s): 1.48 vs. 1.43; time between impact and touchdown (s): 0.46 vs. 0.50; arm movements.
Wang et al., 2022 [113]	Inclusion criteria: ≥60 years. Exclusion criteria: recently (≤6 months) self-reported diagnosed neurological, musculoskeletal, or other systemic disorder.	Community-dwelling elderly: 229 falls (the recovery foot landing posterior to the sliding foot based on the location of heel markers). 569 recoveries.	Subjects walked with their own shoes on a 12 m walkway at their preferred gait speed using a safety harness. Unexpected slips were induced. Kinematic data were collected using an 8-camera optoelectronic motion analysis system (120 Hz). Ground reaction forces were collected (600 Hz).	Stride length (m): 0.41 vs. 0.70; slip distance (m): 0.31 vs. 0.08; slip velocity (m/s): 1.12 vs. −0.34; trunk angle (°): 4.9 vs. −1.5.	
Wang & Bhatt, 2023 [114]	Inclusion criteria: ≥60 years. Exclusion criteria: recently (≤6 months) self-reported diagnosed neurological, musculoskeletal, or other systemic disorder.	Community-dwelling elderly: 61 fallers (the peak moving average force of the load cell over a 1 s period was ≥4.5% of body weight). 56 recoveries.	Subjects walked with their own shoes on a 12 m walkway at their preferred gait speed using a safety harness. Unexpected slips were induced. Kinematic data were collected using an 8-camera optoelectronic motion analysis system (120 Hz). Ground reaction forces were collected (600 Hz). CoM was calculated.	Margin of stability at tripping foot touchdown: 1.39 vs. 0.95; maximum step length (m): 0.11 vs. 0.04; maximum CoM velocity (m/s): 0.59 vs. 0.21; peak trunk angle (°): 23.7 vs. 6.9.	Margin of stability at tripping foot lift-off: 1.03 vs. 0.96; margin of stability at pre-tripping foot touchdown: 2.38 vs. 2.24.

BF—biceps femoris long head; CoM—center of mass; EMG—electromyography; MG—medial gastrocnemius; SD—standard deviation; TA—tibialis anterior; VL—vastus lateralis.

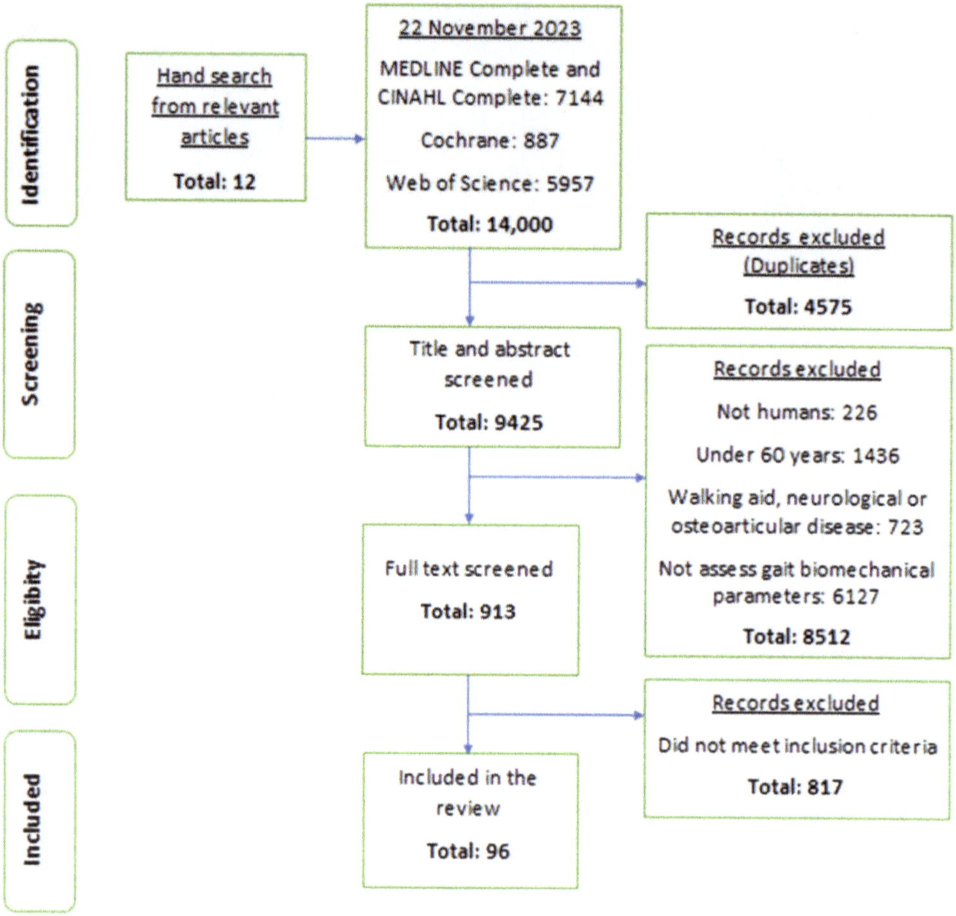

Figure 1. Study flow diagram.

Out of the 86 selected studies that compared elderly fallers and non-fallers, 72 (83.7%) were retrospective, and 12 (14%) were prospective; the remaining two studies compared the elderly who fell from induced falls with the elderly who recovered during unperturbed gait trials. Among the retrospective studies, fifteen evaluated the history of falls during the previous 6 months, one during the previous 10 months, forty-nine during the previous year, two during the previous 2 years, and one during the previous 5 years; four studies did not report this information. Among the prospective studies, two evaluated the occurrence of falls during a 6-month follow-up, eight during a 1-year follow-up, and two during a 2-year follow-up. Additionally, 38 of the studies (44.2%) did not provide a definition of the term "fall".

Regarding the selected retrospective studies, 50 analyzed the subjects' gait on a walkway (forty-four during level ground, one during unleveled ground, and five during gait initiation); 19 on a treadmill (18 during level gait and 1 during gait initiation); and 3 during real-life scenarios. Concerning the prospective studies, 11 analyzed the subjects' level gait on a walkway and 1 on a treadmill.

The elderly's gait was analyzed on a walkway in all studies that induced falls during their methodological set-up (three studies induced trips and nine induced slips).

3.2. Risk of Bias Assessment

Out of the 96 studies, 5 studies had a global classification of moderate, and 91 studies had a global classification of weak (Table 3). Thus, 77 out of the 96 studies were classified as weak regarding the selection bias domain because the subjects were not representative of the study population, i.e., the samples were convenience samples; the remaining 19 studies were classified as moderate because the sample was representative of the population and the studies were completed by 80–100% of the initially included subjects. Regarding the study design domain, the 96 studies were classified as weak since their study design was cross-sectional and the subject selection was not randomized. Relating to the confounders domain, 78 studies were classified as weak because the potential confounders were not shown, and 18 studies were classified as strong because the potential confounders were controlled. Concerning the blinding domain, 24 studies were classified as strong because the investigators were blinded to the status of the subjects, and the subjects were also blinded to the research question, while 72 studies were classified as moderate. Regarding the data collection methods domain, two studies were classified as weak because those methods were not reliable, or the validity and reliability of the instruments were not shown; the remaining 94 studies were classified as strong because it was shown that the instruments were valid and reliable. With respect to the withdrawals and dropouts domain, 84 studies were classified as moderate because the studies were retrospective; in this component, only 12 studies were classified as strong because the percentage of subjects that completed the study was 80% or more.

Table 3. Methodological quality evaluation of the included studies using the Quality Assessment Tool for Quantitative Studies.

Study	Selection Bias	Study Design	Confounders	Blinding	Data Collection Methods	Withdrawals and Dropouts	Global
Heitmann et al., 1989 [20]	3	3	3	2	3	2	3
Gehlsen & Whaley, 1990 [21]	3	3	3	2	1	2	3
Feltner et al., 1994 [22]	3	3	3	2	1	2	3
Wolfson et al., 1995 [23]	3	3	3	2	3	2	3
Maki, 1997 [24]	3	3	1	1	1	1	3
Lee & Kerrigan, 1999 [25]	3	3	3	2	1	2	3
Nelson et al., 1999 [26]	3	3	3	2	1	2	3
Pavol et al., 1999 [105]	3	3	1	1	1	2	3
Wall et al., 2000 [27]	3	3	3	2	3	2	3
Hausdorff et al., 2001 [28]	3	3	3	1	1	1	3
Kerrigan et al., 2000 [29]	3	3	3	2	1	2	3
Kerrigan et al., 2001 [30]	3	3	3	2	1	2	3
Pavol et al., 2001 [107]	3	3	1	1	1	2	3
Kemoun et al., 2002 [31]	3	3	3	1	1	1	3
Auvinet et al., 2003 [32]	3	3	3	2	1	2	3
Mbourou et al., 2003 [33]	3	3	3	2	1	2	3
Lockhart et al., 2003 [106]	3	3	3	1	1	2	3
Chiba et al., 2005 [34]	3	3	1	2	1	2	3
Pijnappels et al., 2005 [103]	3	3	3	1	1	2	3
Barak et al., 2006 [35]	3	3	3	2	1	2	3
Toulotte et al., 2006 [36]	3	3	1	2	1	2	3
Espy & Pai, 2007 [108]	3	3	3	1	1	2	3
Karmakar et al., 2007 [37]	3	3	3	2	1	2	3
Newstead et al., 2007 [39]	3	3	1	2	1	2	3
Barrett et al., 2008 [40]	3	3	3	2	1	2	3
Khandoker et al., 2008 [41]	3	3	3	2	1	2	3

Table 3. *Cont.*

Study	Selection Bias	Study Design	Confounders	Blinding	Data Collection Methods	Withdrawals and Dropouts	Global
Khandoker et al., 2008 [42]	3	3	3	2	1	2	3
Lockhart & Liu, 2008 [43]	3	3	3	2	1	2	3
Verghese et al., 2009 [14]	2	3	1	1	1	1	2
Greany & Di Fabio, 2010 [44]	3	3	3	2	1	2	3
Greene et al., 2010 [45]	2	3	3	2	1	2	3
Mickle et al., 2010 [46]	2	3	1	1	1	1	3
Bhatt et al., 2011 [104]	3	3	3	1	1	2	3
Kirkwood et al., 2011 [47]	3	3	3	2	1	2	3
Lázaro et al., 2011 [48]	2	3	3	2	3	2	3
Lugade et al., 2011 [49]	3	3	3	2	1	2	3
Panzer et al., 2011 [50]	3	3	3	2	3	2	3
Scanaill et al., 2011 [51]	2	3	3	2	1	2	3
Karmakar et al., 2012 [38]	3	3	3	2	1	2	3
Uemura et al., 2012 [52]	3	3	1	2	1	2	3
Chen & Chou, 2013 [53]	3	3	3	2	1	2	3
Chen et al., 2013 [54]	3	3	3	2	1	2	3
Chiu & Chou, 2013 [55]	3	3	3	2	1	2	3
Fritz et al., 2013 [56]	3	3	3	2	1	2	3
Marques et al., 2013 [58]	3	1	3	2	1	2	3
Marques et al., 2013 [59]	3	1	3	2	1	2	3
Weiss et al., 2013 [57]	3	3	1	2	1	2	3
Ayoubi et al., 2014 [60]	2	3	3	2	1	2	3
Barelle et al., 2014 [61]	3	3	3	2	1	2	3
Iwata et al., 2014 [62]	3	3	3	2	1	2	3
Kobayashi et al., 2014 [63]	3	3	3	2	1	2	3
König et al., 2014 [64]	3	3	1	2	1	2	3
Mignardot et al., 2014 [65]	2	3	1	1	1	1	2
Yang & Pai, 2014 [109]	2	3	3	1	1	2	3
Cebolla et al., 2015 [66]	3	3	3	2	1	2	3
MacAulay et al., 2015 [67]	2	3	3	2	1	2	3
Rispens et al., 2015 [68]	3	3	3	2	1	2	3
Wright et al., 2015 [69]	3	3	3	2	1	2	3
Bounyong et al., 2016 [70]	3	3	3	2	1	2	3
Fujimoto & Chou, 2016 [71]	3	3	3	2	1	2	3
Howcroft et al., 2016 [73]	2	3	3	2	1	2	3
Ihlen et al., 2016 [72]	3	3	3	2	1	2	3
Rinaldi et al., 2016 [74]	3	3	3	2	1	2	3
Sawers et al., 2016 [110]	3	3	3	1	1	2	3
Bizovska et al., 2017 [76]	2	3	3	1	1	1	3
de Melker Worms et al., 2017 [77]	3	3	3	2	1	2	3
de Melker Worms et al., 2017 [78]	3	3	3	2	1	2	3
Marques et al., 2017 [79]	3	3	1	2	1	2	3
Júnior et al., 2017 [80]	3	3	3	2	1	2	3
Rinaldi et al., 2017 [75]	3	3	3	2	1	2	3
Svoboda et al., 2017 [81]	2	3	3	2	1	2	3
Allen & Franz, 2018 [82]	3	3	3	2	1	2	3
Benson et al., 2018 [83]	3	3	3	2	1	2	3
Howcroft et al., 2018 [86]	3	3	3	1	1	1	3
Kwon et al., 2018 [84]	3	3	3	2	1	2	3
Marques et al., 2018 [85]	3	3	1	2	1	2	3
Qiao et al., 2018 [88]	3	3	3	2	1	2	3
Sawers & Bhatt, 2018 [111]	3	3	3	1	1	2	3

Table 3. *Cont.*

Study	Selection Bias	Study Design	Confounders	Blinding	Data Collection Methods	Withdrawals and Dropouts	Global
Thompson et al., 2018 [87]	3	3	3	2	1	2	3
Watanabe et al., 2018 [89]	3	3	3	2	1	2	3
Bueno et al., 2019 [90]	3	3	3	2	1	2	3
Gillian et al., 2019 [91]	2	3	1	1	1	1	2
Mak et al., 2019 [92]	2	3	3	2	1	2	3
Yamagata et al., 2019 [93]	3	3	3	1	1	2	3
Yamagata et al., 2019 [94]	3	3	3	1	1	1	3
Gonzalez et al., 2020 [95]	3	3	3	2	1	1	3
Pol et al., 2021 [97]	2	3	1	2	1	2	2
Sadeghi et al., 2021 [98]	3	3	3	2	1	2	3
Yamagata et al., 2021 [96]	3	3	3	1	1	1	3
Bruijn et al., 2022 [112]	3	3	3	1	1	2	3
Figueiredo et al., 2022 [99]	3	3	1	2	1	2	3
Nascimento et al., 2022 [100]	2	3	1	2	3	2	2
Wang et al., 2022 [113]	1	3	3	1	1	2	3
Yoshida et al., 2022 [101]	2	3	3	2	1	2	3
Baba et al., 2023 [102]	3	3	3	2	1	2	3
Wang & Bhatt, 2023 [114]	1	3	3	1	1	2	3

1–strong methodological quality; 2—moderate methodological quality; 3—weak methodological quality.

3.3. *Gait Spatiotemporal Parameters*

The spatiotemporal parameters analyzed among the studies comprised gait speed; cadence; stride and step length; stride and step time; stride and step width; stance phase; swing phase; single support phase; double support phase; and base of support.

3.3.1. Gait Speed

Gait speed was the parameter most analyzed, namely in 50 studies that compared fallers and non-fallers. Regarding these studies, 29 reported the fallers' gait speed was lower than non-fallers [14,23,25–27,30–32,34,37,39,43,48–50,55–57,60,63,68,75,84,85,91,93,98,100,102]. Although another 17 studies observed lower values of gait speed in fallers, no statistically significant differences were yielded [22,24,28,36,58,60–62,66,76,80–82,86,89,91,94]. No study reported higher values of gait speed in fallers.

Four studies analyzed gait speed variability using linear measures: one used the coefficient of variation and standard deviation [81], while three only used the standard deviation [24,63,68]. Of these studies, three reported higher values in fallers [24,68,81], while one reported no differences between fallers and non-fallers [63].

Three studies analyzed gait speed during induced slips [110–112] and two during induced trips [103,105]. Four studies reported no differences between falls and recoveries [103,110–112], while only one reported that the gait speed of the elderly who fell was higher than the elderly who recovered from an induced trip [105].

3.3.2. Cadence

Cadence was evaluated in 22 studies that compared fallers with non-fallers. Among these, six studies reported the fallers' cadence was lower [25,32,39,45,68,98]. Although another 10 studies observed lower values of cadence in fallers, no statistically significant difference was yielded [30,31,36,55,61,86,89,94,100,102]. No study reported higher values of cadence in fallers.

One study analyzed cadence variability (using the standard deviation) and observed no differences between fallers and non-fallers [68].

3.3.3. Stride and Step Length

Stride or step length was analyzed in 39 studies that compared fallers and non-fallers. Nineteen studies reported the fallers' stride or step length was lower [23,25,30,32–35,39,43,51,53,56,67,84,85,91,94,98,100]. Although another 13 studies observed lower values of stride or step length in fallers, no statistically significant difference was yielded [21,22,24,31,36,44,61,63,66,77,80,81,102]. No study reported higher values of stride or step length in fallers.

Stride or step length variability was studied using the coefficient of variation [14,51,77,81] and standard deviation [24,33,63,84,87]. Three studies reported that fallers yielded higher values [14,24,33], while six reported no differences [51,63,77,81,84,87].

Step or stride lengths were also analyzed in studies that induced slips [109,111,113] and trips [103,105,107] during their methodological set-up. Of these, one study reported the stride and step length of the elderly who fell were higher than the elderly who recovered from the induced trip [105], while two studies reported the stride and step length of the elderly who fell were lower than the subjects who recovered from the induced trips [107] and induced slips [113]. The other four studies reported no differences between falls and recoveries [103,109,111].

3.3.4. Stride and Step Time

Stride and step time were evaluated in 37 studies that compared fallers and non-fallers. Nine studies reported the fallers' stride or step time was higher [26,34,40,45,57,68,74,85,98]. Although another 10 studies reported higher values of stride or step time in fallers, no statistically significant differences were yielded [22,24,31,36,43,44,51,65,84,86]. Only one study reported lower values of stride time in fallers [47].

Stride and step time variability were analyzed using the coefficient of variation [45,51,56,60,73,79–81,86] and standard deviation [24,28,40,68,79,84,85]. Six studies reported that fallers yielded higher values [28,40,60,68,79,84], while 10 reported no differences [24,45,51,56,60,73,80,81,85,86].

Step or stride time was also analyzed in studies that induced trips during their methodological set-up [105,107]. These two studies reported the step time of the elderly who fell was lower than the elderly who recovered from induced trips.

3.3.5. Stride and Step Width

Stride and step width were analyzed in 18 studies that compared fallers and non-fallers. Three studies showed that fallers' stride or step width was higher [21,51,74], while one study observed lower values of step width in fallers [51]. The other studies showed no differences between fallers and non-fallers [20–22,24,53,59,63,77,78,81,87,93,94,98,101].

Stride and step width variability was evaluated using the coefficient of variation [51,77,78,81] and standard deviation [20,24,63,87]. All studies reported no differences between fallers and non-fallers.

Step width was also analyzed in studies that induced slips [108] and trips [105] during their methodological set-up. Both reported no differences between fallers and those who recovered [105]. Step width variability was also evaluated (using standard deviation) in one study [109], which verified higher values in fallers.

3.3.6. Stance Phase

The stance phase was evaluated in 13 studies that compared fallers and non-fallers. Four studies reported the fallers' stance phase was higher [40,55,84,85]. The other studies showed no differences between fallers and non-fallers [45,47,51,56,63,73,77,86,102].

The stance phase variability was evaluated using the coefficient of variation [45,51,73,77,79] and standard deviation [40,63,85]. Five studies reported that fallers yielded higher values [40,63,77,79,85], while three reported no differences [45,51,73].

The stance phase was also evaluated in one study that induced trips during its methodological set-up [103], which showed no differences between fallers and those who recovered.

3.3.7. Swing Phase

The swing phase was analyzed in 18 studies that compared fallers and non-fallers. Five studies observed the fallers' swing phase was lower than non-fallers [14,31,55,84,85]. In the other 13 studies, no statistically significant difference was yielded [21,22,40,44,45,47, 51,56,63,73,77,86,94].

The swing phase variability was studied using the coefficient of variation [14,45,51,73,77] and standard deviation [28,40,63,79,85]. Three studies reported that fallers yielded higher values [14,28,79], while seven reported no differences [40,45,51,63,73,77,85].

The swing phase was analyzed in one study that induced trips during its methodological set-up [103], which reported no differences between fallers and those who recovered.

3.3.8. Single Support Phase

The single support phase was evaluated in 10 studies that compared fallers and non-fallers. The fallers' single support phase was lower than non-fallers in two studies [39,55]. The other eight studies showed no differences between fallers and non-fallers [21,22,31,36, 45,51,80,84].

The single support phase variability was also analyzed in two studies using the coefficient of variation [45,51], which observed no differences between fallers and non-fallers.

3.3.9. Double Support Phase

The double support phase was analyzed in 18 studies that compared fallers and non-fallers. Of these, nine studies reported the fallers' double support phase was higher [14,26, 31,33,39,55,84,85,98], while one observed exactly the contrary [45]. The other eight studies showed no difference between fallers and non-fallers [21,24,44,51,56,73,102,103].

The double support phase variability was studied using the standard deviation [24] and coefficient of variation [45,51]. One study reported that fallers yielded higher values [24], while two reported no differences [45,51].

The double support phase was also evaluated in one study that induced trips during its methodological set-up [103], which reported no differences between falls and recoveries.

3.3.10. Base of Support during Gait

The base of support during gait was analyzed in four studies that compared fallers and non-fallers. Of these, two studies reported the fallers' base of support was higher [26,56]. The other two studies showed no differences between fallers and non-fallers [22,49].

The margin of dynamic stability was evaluated in one study that compared fallers and non-fallers [74]; the authors found higher values in fallers. This parameter was also analyzed in one study that induced slips during its methodological set-up [109]. In this study, the authors found higher values in fallers.

3.3.11. Others Parameters

The time of toe-off occurrence (% of the gait stride) was analyzed in one study that compared fallers and non-fallers [89]. This study reported no differences between fallers and non-fallers regarding this parameter.

3.4. Kinematic Parameters

The kinematics parameters analyzed among the studies comprised the following: minimum foot/toe clearance; center of mass (CoM); center of pressure (CoP); head, trunk, pelvis, and lower limb kinematics; and slip kinematic parameters.

3.4.1. Minimum Foot/Toe Clearance

The minimum foot/toe clearance was analyzed in nine studies that compared fallers and non-fallers. Two studies reported the fallers' minimum foot/toe clearance was lower [34,66]. Nonetheless, three studies reported contrary results, i.e., fallers' minimum foot/toe clearance was higher [37,42,89]. On the other hand, four studies reported no differences between fallers and non-fallers [21,41,63,91].

The minimum foot/toe clearance variability was studied using linear measures (i.e., coefficient of variation [34,91] and standard deviation [37,63,91]) and nonlinear measures (i.e., approximate entropy [38,41], sample entropy [38], wavelet-based multiscale exponent [42], detrended fluctuation analysis exponent [42], and Poincaré plot indexes [41,42]). Four studies reported that fallers yielded higher variability [34,37,38,41,42], while two studies reported no differences between fallers and non-fallers [63,91].

3.4.2. CoM

Differences between fallers and non-fallers regarding CoM position were found in two studies [71,93], while one showed no differences [49]. One study found no differences regarding CoM displacement [96], while another found differences regarding CoM lateral sway [35]. Three studies observed lower values of the fallers' CoM velocity during gait [53,71,74], namely AP velocity [53,74], while one found no differences [105]. Regarding CoM acceleration, one study found lower values in fallers [71].

CoM variability was analyzed in one study using the variance [94]. In this study, higher values were found in fallers regarding the variability in the CoM vertical displacement; however, concerning the variability in the CoM mediolateral (ML) displacement, no differences were observed between fallers and non-fallers. On the other hand, one study using the local divergence exponent found higher values in fallers, i.e., higher variability [77].

CoM was also analyzed in one study that induced trips during its methodological set-up, which reported the CoM position and velocity of the elderly who fell were not different than the elderly who recovered from the induced trip [108].

The dynamic stability of CoM was analyzed in one study that induced slips during its methodological set-up (dynamic stability is the relative motion state between CoM and the base of support). This study found no differences between the elderly who fell and the elderly who recovered from the induced slip [109]. The dynamic stability was also analyzed in three studies that induced slips during their methodological set-up. Of this, one study reported the fallers' dynamic stability was higher [104]. The other two studies showed no differences [110,111].

3.4.3. CoP Kinematics

One study reported that the fallers' CoP excursion index was lower than non-fallers [97]. Moreover, one study found higher CoP ML displacement in fallers [101]. On the other hand, four studies reported the CoP AP displacement and/or velocity presented no differences between fallers and non-fallers [73,86,97,101].

The variability in the CoP AP and ML displacements was evaluated in two studies using the standard deviation [81] and coefficient of variation [73]. Their authors reported no differences between fallers and non-fallers.

3.4.4. CoM–CoP Relation

Two studies that compared fallers and non-fallers analyzed the CoM–CoP relation. Their results are contradictory. While one study reported the fallers' CoM–CoP AP distance was lower [49], the other observed higher values [69]. On the other hand, CoM–CoP ML distance presented no difference between fallers and non-fallers in one study [49].

3.4.5. Head, Trunk, and Pelvis Linear Kinematics

Trunk linear kinematics were evaluated in one study that compared fallers and non-fallers [34]; they found higher maximal ML displacement of the trunk center in fallers.

One study used the refined composite multiscale entropy and the refined multiscale permutation entropy regarding lower back velocity and acceleration [72]; they found higher complexity in fallers. The computed multiscale entropy and the Shannon entropy were also used to analyze the complexity of the trunk AP and ML displacement [76]; data pointed out the inability of the multiscale entropy to distinguish fallers and non-fallers, whereas Shannon entropy seemed to be sufficient in fall risk prediction. On the other hand, fallers presented higher variability in the lower back vertical axis and lower variability in the lower back ML axis [57].

One study used the short-term exponents of the trunk ML, AP, and vertical displacement to analyze gait variability. No differences were yielded between fallers and non-fallers [95]. One study that during their methodological set-up induced slips [109] also analyzed the variability in the trunk through nonlinear measures, i.e., using the maximum Lyapunov exponent and Floquet multiplier. Their authors found no differences between fallers and those who recovered.

The maximum Lyapunov exponent was also used in two studies in order to evaluate the gait variability. Contradictory results were found in these two studies, i.e., one found higher variability in the right anterior superior iliac spine in fallers [43], while the other did not find differences between fallers and non-fallers regarding the head and pelvis [73]. One of these studies also analyzed variability using the ratio of even to odd harmonics, having found differences between fallers and non-fallers regarding the pelvis [73].

3.4.6. Lower Limb Linear Kinematics

Two studies analyzed the foot velocity and heel vertical velocity at heel strike [44,66]. They found no differences between fallers and non-fallers.

One study used the fast Fourier transform first quartile on the shank displacement [86]; they found higher variability in fallers.

The hip horizontal displacement [103], the hip–ankle distance, the obstacle–ankle distance, the ankle horizontal velocity, and the hip vertical velocity [107] were analyzed in two studies that induced trips during their methodological set-up. No differences were found between fallers and those who recovered.

The hip height was analyzed in two studies that induced trips during their methodological set-up [103,107]. Of this, one study reported the fallers' hip height at ground foot contact was lower [107]. The other study showed no differences between the elderly who fell or recovered from the induced trips [103].

3.4.7. Slip Kinematics Parameters

The slip distance was analyzed in three studies that induced slips during their methodological set-up [110,111,113]. Of these, two studies reported the fallers' slip distance was higher [110,113]. The other study showed no differences [111]. The peak slip velocity was also evaluated in these three studies. Of these, one study reported the fallers' peak slip velocity was higher [113]; the other two studies showed no differences [110,111]. One of these three studies also assessed the slip duration [111]. In this study, no differences were found between fallers and recoverers.

3.5. Angular Kinematic Parameters

The angular kinematics analyzed among the selected studies comprised the lower limb joints (hip, knee, and ankle), foot progression angle, and foot angle with the ground, trunk, pelvis, thigh, and shank.

3.5.1. Hip

Five studies reported differences between fallers and non-fallers regarding hip angular position or displacement [29,31,35,58,63]. On the other hand, eight studies found no differences regarding hip angular position or displacement [21,22,29,31,61,82,98,102].

Fallers exhibited greater variability in the hip in the frontal plane during the entire stance phase [63].

Hip angular position was also analyzed in two studies that induced trips [107] and slips [111] during their methodological set-up. The first one found differences between fallers and those who recovered, while the other did not.

3.5.2. Knee

Knee kinematics were evaluated in 10 studies that compared fallers and non-fallers. These studies reported no differences regarding knee angular position or displacement [21,22,30,31,61,82,83,90,98,102].

Fallers exhibited greater variability in the knee during the entire swing phase [63].

Knee angular position was also analyzed in one study that induced slips during its methodological set-up [111]. Data yielded higher values of knee flexion in fallers but no differences regarding knee extension.

3.5.3. Ankle

Ankle angular position or displacement yielded no differences between fallers and non-fallers in nine studies [21,22,30,58,61,82,90,98,102]. On the other hand, differences between fallers and non-fallers regarding ankle kinematics were found in three studies [31,35,63].

Fallers exhibited greater variability in the ankle in the frontal plane during the entire stance phase [63].

Ankle angular position was analyzed in one study that induced slips during its methodological set-up [111]. No differences were found between fallers and non-fallers.

3.5.4. Foot Progression Angle

Foot progression angle was analyzed in three studies that compared fallers and non-fallers. These studies supported no differences between fallers and non-fallers [84,90,94].

Foot progression angle variability was also studied using the standard deviation [93,94]. No differences were found between fallers and non-fallers.

3.5.5. Foot Angle with Ground

Differences between fallers and non-fallers regarding the foot angle with the ground were found in two studies [34,102]. While one study observed higher values in fallers [34], the other found lower values [102].

The variability in the maximum foot angle with the ground was also studied using the coefficient of variation [34]. In this study, higher values were found in fallers.

3.5.6. Trunk

Trunk angular position was analyzed in one study that compared fallers and non-fallers; no differences were found [54].

Trunk angular position was also evaluated in three studies that induced trips [105,107] and slips [114] during their methodological set-up. Two of these studies observed differences between fallers and non-fallers [107,114], while the other did not [105].

3.5.7. Pelvis

Pelvis angular position was analyzed in three studies that compared fallers and non-fallers. All studies reported no differences between fallers and non-fallers [30,93,94].

The variability in pelvis angular position was studied using the standard deviation [93,94]; however, no differences were yielded between fallers and non-fallers.

3.5.8. Thigh

Thigh angular position was analyzed in two studies that compared fallers and non-fallers [93,94], which reported no differences between these groups.

The variability in thigh angular position was studied using the standard deviation [93,94]. No differences between fallers and non-fallers were found.

3.5.9. Shank

Three studies that compared fallers and non-fallers reported differences between fallers and non-fallers regarding shank angular position [45,93,94].

The variability in the shank angular position was studied using the standard deviation [93,94]. Differences between fallers and non-fallers were found, with higher values in fallers.

Shank angular position was also analyzed in two studies that induced slips during their methodological set-up [110,111]. No differences between fallers and non-fallers were yielded.

3.5.10. Other Parameters

AP CoM–ankle inclination was evaluated in one study that compared fallers and non-fallers [53]; the authors found higher inclinations in fallers.

The ankle–hip inclination at the time of loading was analyzed in one study that induced trips during its methodological set-up [107]. In this study, differences between fallers and those who recovered were found, i.e., higher inclinations in fallers.

Two studies analyzed the variability in inter-joint coordination. One study used the standard deviation for this purpose [55] and found higher variability in fallers regarding the knee–ankle coordination (during stance and swing phase); however, the variability in hip–knee coordination yielded no differences between fallers and non-fallers [55]. In addition, another study reported lower variability in the lower limb coordination in fallers, indicating an inconsistency in neuromuscular control [98].

3.6. Kinetic Parameters

3.6.1. Ground Reaction Force

The ground reaction force was analyzed in five studies that compared fallers and non-fallers [54,69,73,84,86]. The peak braking force and the peak propulsive force (calculated from the AP component) presented no differences between fallers and non-fallers [69,84]. On the other hand, differences regarding braking force and propulsive force were found in another study [54]. On the other hand, the impulse during the gait cycle presented no differences in two studies [73,86].

Ground reaction force was also evaluated in one study that induced slips during its methodological set-up [101]. This study reported the fallers' coefficient of friction (horizontal ground reaction force/vertical ground reaction force after heel contact) was lower than in those who recovered.

3.6.2. Plantar Pressure

Peak plantar pressure and total pressure–time integral presented higher values in fallers [46,97]. On the other hand, the pressure–time integral regarding different foot regions (medial and lateral heel, medial and lateral midfoot, central and lateral forefoot, hallux, second and lateral toe) showed no difference between fallers and non-fallers; only the pressure–time integral of medial forefoot yielded higher values in fallers [98].

3.7. Dynamic Parameters

3.7.1. Hip Moment

The hip moment was analyzed in four studies that compared fallers and non-fallers [25, 30,31,54]. Three studies point out differences, namely in the sagittal plane [25,30,31]. Only one study reported no differences between fallers and non-fallers [54]. Moreover, one study also evaluated other planes of movement [30]; they found differences regarding the hip adduction moment but no differences concerning hip abduction, external, and internal moments.

The hip moment was analyzed in one study that induced trips during its methodological set-up [103], which reported no differences between fallers and those who recovered.

3.7.2. Knee Moment

The knee moment was analyzed in four studies that compared fallers and non-fallers. Two studies yielded differences between fallers and non-fallers [25,30], while the other two reported no differences [31,54].

The knee moment was also evaluated in one study that induced trips during its methodological set-up [103]. In this study, no differences were yielded.

3.7.3. Ankle Moment

The ankle moment was evaluated in four studies that compared fallers and non-fallers. Two studies yielded differences between fallers and non-fallers [25,54], while the other two reported no differences [30,31].

Ankle moment was also analyzed in one study that induced trips during its methodological set-up [103]. The authors reported no differences between fallers and non-fallers.

3.7.4. Hip, Knee, and Ankle Power Absorption and Generation

One study found differences between fallers and non-fallers regarding hip power absorption and generation, namely with lower values in fallers [30]. However, another two studies reported no differences between fallers and non-fallers [25,31].

Two studies reported the knee power absorption of fallers was lower than non-fallers; however, the knee power peak and power generation presented no differences [25,30].

Ankle power absorption and generation were analyzed in three studies that compared fallers and non-fallers [25,30,31]. One study reported the ankle power generation of fallers was lower than non-fallers [30], while another observed higher values of ankle power absorption in fallers [25]. On the other hand, another study reported the ankle power peak presented no differences between fallers and non-fallers [31].

3.7.5. Other Parameters

One study found differences between fallers and non-fallers regarding total CoM kinetic energy (at swing-off), namely with lower values in fallers [53].

Angular momentum was also evaluated in one study that induced trips during its methodological set-up [103]. This study reported the fallers' angular momentum at push-off and total momentum are predictors of falls.

3.8. EMG Parameters

3.8.1. Muscle Activity

Muscle activity was analyzed in 10 studies using EMG measures. Eight studies compared elderly fallers and non-fallers [47,59,70,82,88,89,92,94], while two studies compared the elderly who fell with the elderly who recovered from induced slips [110,111].

The fallers' internal oblique activity before heel contact was lower than non-fallers. On the other hand, the same study observed no significant differences in internal oblique activity at the initial stance, final stance, and after toe-off [59].

The gluteus maximus was analyzed in one study that compared fallers and non-fallers. This study reported the fallers' gluteus maximus activity at the final stance is higher than non-fallers. On the other hand, no significant differences in gluteus maximus activity were observed at the initial stance, before heel contact, and after toe-off [59].

The fallers' biceps femoris activity at initial stance and before heel contact is higher than non-fallers. On the other hand, no significant differences in biceps femoris activity were observed at the final stance and after toe-off [59].

The biceps femoris long head was analyzed in one study that induced slips during its methodological set-up. Their authors pointed out the higher onset latencies of fallers. On the other hand, no differences were observed regarding the onset latencies of the nonslip leg and the peak magnitude of the slip and nonslip leg [110].

The gastrocnemius was analyzed in two studies that compared fallers and non-fallers [47,59]. One of these studies reported the fallers' gastrocnemius activity during the stance phase was lower than non-fallers [47]. On the other hand, no differences were observed in gastrocnemius activity at the initial stance, before heel contact, and after toe-off [47,59].

The medial gastrocnemius yielded differences in onset latencies and peak magnitude when the elderly who fell and the elderly who recovered from an induced slip were compared [110].

The vastus lateralis was analyzed in one study that induced slips during its methodological set-up. The authors reported the fallers' onset latencies of the slip leg were higher than in those who recovered. On the other hand, no differences in onset latencies of the nonslip leg and peak magnitude of the slip and nonslip leg were reported [110].

No differences between fallers and non-fallers were found regarding soleus activity and onset latency [47], tibialis anterior activity [59,110], latency [47,110], peak magnitude [110], multifidus activity [59,110], or rectus femoris activity [59,110].

The variability in central locus activation of the rectus femoris was studied using the coefficient variation [98]. No differences between fallers and non-fallers were found.

3.8.2. Muscle Synergies and Co-Contraction

The co-contraction index (between tibialis anterior and gastrocnemius) and lower limb muscle co-contractions of fallers were higher than non-fallers [59,70].

The muscle synergies were analyzed in two studies that compared fallers and non-fallers. One study reported the fallers' muscle synergies were lower than non-fallers [82]. On the other hand, fallers' kinematic synergy during the early and late swings was higher than non-fallers [94].

The muscle synergies were also evaluated in two studies that induced slips during their methodological set-up. In both studies, it was reported that the fallers' muscle synergies were lower than in those who recovered [110,111].

3.9. *Gait Symmetry and Gait Smoothness*

Two studies found differences between fallers and non-fallers regarding gait symmetry, which expressed the similarity of craniocaudal movements on the left and the right independently from fluctuations in successive craniocaudal movements of each limb [32,68].

The two studies that analyzed gait smoothness—a quality that reflects the continuousness or non-intermittency of walking [68,99]—found different results, i.e., one study reported that gait smoothness was associated with the number of falls [68], while the other did not find differences between fallers and non-fallers [99].

4. Discussion

The aim of the present study was to conduct a systematic review to identify and describe the gait biomechanical parameters related to falls in the elderly population. According to the results of this systematic review, the gait spatiotemporal parameters were the most analyzed data when elderly fallers and non-fallers were compared, especially the gait speed. The majority of the selected studies for this systematic review reported lower gait speed in elderly fallers, pointing out that this can be a gait biomechanical parameter that differentiates elderly fallers from non-fallers. This lower gait speed in fallers may be the result of reduced functional capacity, fear of falling, or both. Concerning functional capacity, the data compiled in this systematic review may provide some clues: differences between elderly fallers and non-fallers were found regarding lower limb muscular activity and lower limb joints biomechanics (i.e., joints moments and powers), although the number of studies on these topics has been scarce. On the other hand, gait speed has been associated with a fear of falling [24,52,60]. In this way, 63.4% of the retrospective studies (i.e., studies in which fallers had already suffered a fall at the time of gait assessment) that analyzed gait speed reported a lower gait speed in fallers, while only 33.3% of the prospective studies (i.e., studies in which fallers had not suffered a fall at the time of gait assessment) reported a lower gait speed in fallers. These numbers also suggest an effect of fear of falling again on the elderly gait, namely on gait speed. Other parameters were also referenced as related to the fear of falling, such as the stride/step length and the double support phase [24]. According to the data in this systematic review, these parameters have also presented the ability to differentiate elderly fallers from non-fallers, i.e., fallers tended to present a reduced stride/step length and an increased double support phase. Therefore, these data point to the importance of interventions in the elderly who restore and improve their functional capacity and self-confidence during activities of daily living.

Gait speed is dependent on cadence and step length [115]. According to data from this systematic review, the lower gait speed shown by fallers seems to be more sensitive to a reduction in step length than in cadence, although several studies have also found a lower cadence in fallers. On the other hand, stride and step time were other spatiotemporal parameters associated with gait speed [115]; according to our data, these parameters showed lower ability than gait speed to differentiate fallers from non-fallers. Finally, stride/step width and foot progression angle were gait biomechanical parameters that clearly did not differentiate fallers from non-fallers.

Tripping is one of the most frequent causes of falls in the elderly [9], while the minimum foot/toe clearance has been a gait biomechanical parameter associated with trips [11]. Regarding this parameter, our data point to contradictory and non-differentiating results, suggesting the minimum foot/toe clearance may not be a consistent differentiator between fallers and non-fallers. In contrast, the minimum foot/toe clearance variability, using both linear and nonlinear measures, appears as a potential parameter associated with a history of falls in the elderly. This is in line with a previous systematic review [11], which concluded that higher minimum foot clearance variability may contribute to an increased risk of trips.

The minimum foot/toe clearance Is sensitive to alterations in the angular positions of the swing limb joints, i.e., hip, knee, and ankle [116]. Previous research pointed to a higher sensitivity to the ankle angular position, then to the hip angular position, and lastly to the knee angular position [116–118]. Thus, the elderly can adjust minimum foot/toe clearance by controlling these joint angles. In this way, the analysis of lower limb joint kinematics among fallers and non-fallers is an important issue. Although it was not transversal to all studies selected for this systematic review, some of them identified differences between fallers and non-fallers regarding hip and ankle angular position or displacement. Thus, the need for interventions that improve the motor control of the lower limb joints is a very important aspect to explore. In this regard, a program of proprioceptive and functional strength exercises seems to be a good solution for improving motor control of the lower limb joints [119,120]. On the other hand, knee kinematics do not seem to have the ability to differentiate fallers from non-fallers.

Studies involving gait with induced stability perturbations were relatively scarce and analyzed a large disparity of variables (with low frequency in the various studies). Among the studies that induced trips during their methodological set-up, a lowering strategy, characterized by a faster gait speed and delayed support limb loading, was linked to falls during a step [107]. On the other hand, elevating strategy falls were marked by an accelerated gait speed and excessive lumbar flexion. Moreover, fallers exhibited an insufficient reduction in angular momentum during push-off, improper recovery limb placement, and reduced rates of moment generation in support limb joints. Due to the fact that research on this issue is scarce, the extraction of conclusions is largely limited.

Slip falls are a growing health concern for the elderly [113]. Heel velocity and foot angle with the ground at heel strike were gait biomechanical parameters associated with slip falls [12,34,102]. According to data from this systematic review, the literature is scarce regarding these parameters. Moreover, foot angle with the ground at the heel strike revealed contradictory results, and heel velocity at the heel strike showed no differences between fallers and non-fallers. Moreover, the studies that induced slips during their methodological set-up assessed other gait biomechanical parameters, such as the coefficient of friction, the slip distance, the peak slip velocity, and the slip duration. Once again, the studies that analyzed these issues are scarce; nonetheless, their data indicated the fallers' coefficient of friction (horizontal ground reaction force/vertical ground reaction force, after heel contact) was lower than in those who recovered. Regarding the other parameters, there seems to be no great ability to differentiate fallers from those who recovered.

Postural stability can be defined as the ability to maintain adequate sustainability of the body along the movement [121]. The CoM and CoP have been used to analyze postural stability in previous studies [122–124]. The literature is scarce concerning the relation between falls and CoM and CoP; however, the contradictory findings across studies emphasize the complexity of assessing postural control and stability during gait, especially regarding the comparison between fallers and non-fallers. Moreover, the analysis of the base of support during gait as well as the margin of dynamic stability also revealed diverse findings, making it impossible to draw clear conclusions.

The variability in the gait biomechanical parameters was studied using linear and nonlinear measures. Overall, fallers tend to exhibit higher variability in the gait pattern. According to previous research [24], this higher gait variability can be linked with lower motor control. As described in a previous paragraph, exercise programs can be good options in order to improve motor control and, as a result, to reduce the risk of falls.

The majority of the included studies were classified as weak in the global assessment, reflecting concerns about their quality. Selection bias was a domain that influenced this classification, with the majority of studies categorized as weak. The dependence on convenience samples has raised concerns regarding the generalizability to broader populations, restricting the external validity of these studies. Additionally, all studies were rated as weak in the study design domain, predominantly attributable to the cross-sectional nature of the investigations and the nonrandomized selection of the subjects. This is not precisely a problem, as our study aimed to compare the gait biomechanical parameters between elderly fallers and non-fallers. Other limitations identified across the various studies include the definition of a fall, the timeframe considered for the fall's occurrence, and the gait assessment on a treadmill. Approximately 55% of the selected studies presented a definition of fall and these definitions were quite similar. However, nearly half of the studies did not provide any explicit definition, which may impact the reliability of the reported results. According to data from this systematic review, approximately 21% of the selected studies conducted the gait assessment on a treadmill. Walking on the treadmill does not reflect everyday gait as it imposes a different gait pattern [125]. In this way, the obtained results may be influenced by this methodological constraint.

Practical and clinical implications arise from this study. In this way, healthcare workers and clinicians must pay attention to some gait biomechanical parameters (i.e., gait speed, stride/step length, and double support phase) when evaluating elderly gait and during interventions that aim to prevent the occurrence of falls.

Some of the gait biomechanical analysis methods used in the selected studies are not considered the gold standard methods to assess gait, i.e., optoelectronic gait analysis systems and force plates [126–128]. Nonetheless, most of the other equipment used in the various studies is presented in the literature as validated and reliable. This heterogeneity observed between the reviewed studies regarding the gait biomechanical analysis methods used may limit our conclusions, contributing to different values for the same parameter. Therefore, this is one of the reasons why no meta-analysis was carried out. Therefore, it is preferable for future investigations to use gold-standard methods to assess gait in the elderly.

Indeed, the scarce literature regarding some parameters limited the ability of this study to yield strong conclusions. In this way, further work is needed to understand the association between the gait biomechanical parameters and falls in the elderly. An example is the need for research that comprises parameters associated with motor control, such as muscular activity. Another aspect that we did not see addressed in the studies selected for this systematic review was joint stability. One parameter used to study joint stability is dynamic joint stiffness [121], which has been used to differentiate fallers and non-fallers in certain clinical populations [129].

5. Conclusions

The results of this systematic review pointed out that the gait speed, stride/step length, and double support phase are biomechanical parameters of gait that play a distinctive role in differentiating fallers from non-fallers. In this way, these are parameters that healthcare workers and clinicians must pay attention to when evaluating elderly gait and during interventions that aim to prevent the occurrence of falls. Elderly fallers also tend to exhibit higher variability; the variability in the minimum foot/toe clearance is an important example due to its relation with trips. Although studies on lower limb muscular activity and joint biomechanics are limited, the available research indicated that differences in these aspects may also be associated with the propensity for falls. However, it is crucial to highlight the complexity of drawing clear conclusions due to the scarcity of literature and contradictory results among studies, namely parameters related to postural stability. Parameters such as minimum foot/toe clearance, step width, and knee kinematics did not demonstrate a discriminative ability between fallers and non-fallers. Therefore, despite advancements in understanding the biomechanics of gait concerning falls, further research is needed at some points to provide a more comprehensive and consistent understanding of these complex relationships.

Author Contributions: Conceptualization, J.A., T.A. and P.A.; methodology, J.S., T.A. and P.A.; validation, J.A., T.A. and P.A.; formal analysis, J.A., T.A. and P.A.; investigation, J.S. and P.A.; resources, J.S. and P.A.; data curation, J.S. and P.A.; writing—original draft preparation, J.S.; writing—review and editing, J.A., T.A. and P.A.; supervision, P.A.; project administration, P.A. All authors have read and agreed to the published version of the manuscript.

Funding: This research received no external funding.

Institutional Review Board Statement: Not applicable.

Informed Consent Statement: Not applicable.

Data Availability Statement: Not applicable.

Conflicts of Interest: The authors declare no conflicts of interest.

References

1. World Health Organization. Falls. Available online: https://www.who.int/news-room/fact-sheets/detail/falls (accessed on 7 October 2021).
2. Stenhagen, M.; Ekström, H.; Nordell, E.; Elmståhl, S. Accidental falls, health-related quality of life and life satisfaction: A prospective study of the general elderly population. *Arch. Gerontol. Geriatr.* **2014**, *58*, 95–100. [CrossRef] [PubMed]
3. Yardley, L.; Smith, H. A prospective study of the relationship between feared consequences of falling and avoidance of activity in community-living older people. *Gerontologist* **2002**, *42*, 17–23. [CrossRef] [PubMed]
4. Ambrose, A.; Cruz, L.; Paul, G. Falls and fractures: A systematic approach to screening and prevention. *Maturitas* **2015**, *82*, 85–93. [CrossRef]
5. World Health Organization. *WHO Global Report on Falls Prevention in Older Age*; World Health Organization: Geneva, Switzerland, 2007.
6. Florence, C.; Bergen, G.; Atherly, A.; Burns, E.; Stevens, J.; Drake, C. Medical costs of fatal and nonfatal falls in older adults. *J. Am. Geriatr. Soc.* **2018**, *66*, 693–698. [CrossRef]
7. American Geriatrics Society; British Geriatrics Society; American Academy of Orthopaedic Surgeons Panel on Falls Prevention. Guideline for the prevention of falls in older persons. *J. Am. Geriatr. Soc.* **2001**, *49*, 664–672. [CrossRef]

8. Rubenstein, L. Falls in older people: Epidemiology, risk factors and strategies for prevention. *Age Ageing* **2006**, *35* (Suppl. S2), ii37–ii41. [CrossRef] [PubMed]
9. Creaby, M.; Cole, M. Gait characteristics and falls in Parkinson's disease: A systematic review and meta-analysis. *Park. Relat. Disord.* **2018**, *57*, 1–8. [CrossRef]
10. Berg, W.; Alessio, H.; Mills, E.; Tong, C. Circumstances and consequences of falls in independent community-dwelling older adults. *Age Ageing* **1997**, *26*, 261–268. [CrossRef]
11. Barrett, R.; Mills, P.; Begg, R. A systematic review of the effect of ageing and falls history on minimum foot clearance characteristics during level walking. *Gait Posture* **2010**, *32*, 429–435. [CrossRef]
12. Aleixo, P.; Abrantes, J. 3D gait analysis in rheumatoid arthritis postmenopausal women with and without falls history. In Proceedings of the IEEE 2015 4th Portuguese Meeting on Bioengineering (ENBENG), Porto, Portugal, 26–28 February 2015; pp. 1–4. [CrossRef]
13. Callisaya, M.; Blizzard, L.; Schmidt, M.; Martin, K.L.; McGinley, J.L.; Sanders, L.M.; Srikanth, V.K. Gait, gait variability and the risk of multiple incident falls in older people: A population-based study. *Age Ageing* **2011**, *40*, 481–487. [CrossRef]
14. Verghese, J.; Holtzer, R.; Lipton, R.; Wang, C. Quantitative gait markers and incident fall risk in older adults. *J. Gerontol. A Biol. Sci. Med. Sci.* **2009**, *64A*, 896–901. [CrossRef]
15. Page, M.; McKenzie, J.; Bossuyt, P.; Boutron, I.; Hoffmann, T.; Mulrow, C.; Shamseer, L.; Tetzlaff, J.; Akl, E.; Brennan, S.; et al. The PRISMA 2020 statement: An updated guideline for reporting systematic reviews. *BMJ* **2021**, *372*, n71. [CrossRef]
16. Schwartz, M.; Rozumalski, A. The gait deviation index: A new comprehensive index of gait pathology. *Gait Posture* **2008**, *28*, 351–357. [CrossRef]
17. Page, M.J.; Moher, D.; Bossuyt, P.; Boutron, I.; Hoffmann, T.; Mulrow, C.; Shamseer, L.; Tetzlaff, J.; Akl, E.; Brennan, S.; et al. PRISMA 2020 explanation and elaboration: Updated guidance and exemplars for reporting systematic reviews. *BMJ* **2021**, *372*, n160. [CrossRef]
18. World Health Organization. Integrated Care for Older People: Guidelines on Community-Level Interventions to Manage Declines in Intrinsic Capacity. Available online: https://apps.who.int/iris/handle/10665/258981 (accessed on 7 October 2021).
19. Effective Public Health Practice Project. Quality Assessment Tool for Quantitative Studies. Available online: https://www.ephpp.ca/quality-assessment-tool-for-quantitative-studies/ (accessed on 7 October 2022).
20. Heitmann, D.; Gossman, M.; Shaddeau, S.; Jackson, J. Balance performance and step width in noninstitutionalized, elderly, female fallers and nonfallers. *Phys. Ther.* **1989**, *69*, 923–931. [CrossRef]
21. Gehlsen, G.; Whaley, M. Falls in the elderly: Part I, gait. *Arch. Phys. Med. Rehabil.* **1990**, *71*, 735–738. [PubMed]
22. Feltner, M.; MacRae, P.; McNitt-Gray, J. Quantitative gait assessment as a predictor of prospective and retrospective falls in community-dwelling older women. *Arch. Phys. Med. Rehabil.* **1994**, *75*, 447–453. [CrossRef] [PubMed]
23. Wolfson, L.; Judge, J.; Whipple, R.; King, M. Strength is a major factor in balance, gait, and the occurrence of falls. *J. Gerontol. A Biol. Sci. Med. Sci.* **1995**, *50A*, 64–67. [CrossRef]
24. Maki, B. Gait changes in older adults: Predictors of falls or indicators of fear? *J. Am. Geriatr. Soc.* **1997**, *45*, 313–320. [CrossRef]
25. Lee, L.; Kerrigan, D. Identification of kinetic differences between fallers and nonfallers in the elderly. *Am. J. Phys. Med. Rehabil.* **1999**, *78*, 242–246. [CrossRef] [PubMed]
26. Nelson, A.; Certo, L.; Lembo, L.; Lopez, D.; Manfredonia, E.; Vanichpong, S.; Zwick, D. The functional ambulation performance of elderly fallers and non-fallers walking at their preferred velocity. *NeuroRehabilitation* **1999**, *13*, 141–146. [CrossRef]
27. Wall, J.; Bell, C.; Campbell, S.; Davis, J. The Timed Get-up-and-Go test revisited: Measurement of the component tasks. *J. Rehabil. Res. Dev.* **2000**, *37*, 109–113. Available online: http://www.ncbi.nlm.nih.gov/pubmed/10847578 (accessed on 22 November 2023). [PubMed]
28. Hausdorff, J.; Rios, D.; Edelberg, H. Gait variability and fall risk in community-living older adults: A 1-year prospective study. *Arch. Phys. Med. Rehabil.* **2001**, *82*, 1050–1056. [CrossRef]
29. Kerrigan, D.; Lee, L.; Nieto, T.; Markman, J.; Collins, J.; Riley, P. Kinetic alterations independent of walking speed in elderly fallers. *Arch. Phys. Med. Rehabil.* **2000**, *81*, 730–735. [CrossRef] [PubMed]
30. Kerrigan, D.; Lee, L.; Collins, J.; Riley, P.; Lipsitz, L. Reduced hip extension during walking: Healthy elderly and fallers versus young adults. *Arch. Phys. Med. Rehabil.* **2001**, *82*, 26–30. [CrossRef]
31. Kemoun, G.; Thoumie, P.; Boisson, D.; Guieu, J. Ankle dorsiflexion delay can predict falls in the elderly. *J. Rehabil. Med.* **2002**, *34*, 278–283. [CrossRef]
32. Auvinet, B.; Berrut, G.; Touzard, C.; Moutel, L.; Collet, N.; Chaleil, D.; Barrey, E. Gait abnormalities in elderly fallers. *J. Aging Phys. Act.* **2003**, *11*, 40–52. [CrossRef]
33. Mbourou, G.; Lajoie, Y.; Teasdale, N. Step length variability at gait initiation in elderly fallers and non-fallers, and young adults. *Gerontology* **2003**, *49*, 21–26. [CrossRef] [PubMed]
34. Chiba, H.; Ebihara, S.; Tomita, N.; Sasaki, H.; Butler, J. Differential gait kinematics between fallers and non-fallers in community-dwelling elderly people. *Geriatr. Gerontol. Int.* **2005**, *5*, 127–134. [CrossRef]
35. Barak, Y.; Wagenaar, R.; Holt, K. Gait characteristics of elderly people with a history of falls: A dynamic approach. *Phys. Ther.* **2006**, *86*, 1501–1510. [CrossRef]
36. Toulotte, C.; Thevenon, A.; Watelain, E.; Fabre, C. Identification of healthy elderly fallers and non-fallers by gait analysis under dual-task conditions. *Clin. Rehabil.* **2006**, *20*, 269–276. [CrossRef] [PubMed]

37. Karmakar, C.; Khandoker, A.; Begg, R.; Palaniswami, M.; Taylor, S. Understanding ageing effects by approximate entropy analysis of gait variability. In Proceedings of the 29th Annual International Conference of the IEEE Engineering in Medicine and Biology Society, Lyon, France, 22–26 August 2007; IEEE: Piscataway, NJ, USA, 2007; pp. 1965–1968. [CrossRef]
38. Karmakar, C.; Khandoker, A.; Begg, R.; Palaniswami, M. Understanding ageing effects using complexity analysis of foot–ground clearance during walking. *Comput. Methods Biomech. Biomed. Engin.* **2013**, *16*, 554–564. [CrossRef]
39. Newstead, A.; Walden, J.; Gitter, A. Gait variables differentiating fallers from nonfallers. *J. Geriatr. Phys. Ther.* **2007**, *30*, 93–101. [CrossRef] [PubMed]
40. Barrett, A.; O'Connor, M.; Culhane, K.; Finucane, A.; ÓLaighin, G. A footswitch evaluation of the gait of elderly fallers with and without a diagnosis of orthostatic hypotension and healthy elderly controls. In Proceedings of the 30th Annual International Conference of the IEEE Engineering in Medicine and Biology Society, Vancouver, BC, Canada, 20–25 August 2008; IEEE: Piscataway, NJ, USA, 2008; pp. 5101–5104. [CrossRef]
41. Khandoker, A.; Palaniswami, M.; Begg, R. A comparative study on approximate entropy measure and poincaré plot indexes of minimum foot clearance variability in the elderly during walking. *J. Neuroeng. Rehabil.* **2008**, *5*, 4. [CrossRef]
42. Khandoker, A.; Taylor, S.; Karmakar, C.; Begg, R.; Palaniswami, M. Investigating scale invariant dynamics in minimum toe clearance variability of the young and elderly during treadmill walking. *IEEE Trans. Neural Syst. Rehabil. Eng.* **2008**, *16*, 380–389. [CrossRef] [PubMed]
43. Lockhart, E.; Liu, J. Differentiating fall-prone and healthy adults using local dynamic stability. *Ergonomics* **2008**, *51*, 1860–1872. [CrossRef]
44. Greany, F.; Di Fabio, R. Models to predict fall history and fall risk for community-dwelling elderly. *Phys. Occup. Ther. Geriatr.* **2010**, *28*, 280–296. [CrossRef]
45. Greene, B.; O'Donovan, A.; Romero-Ortuno, R.; Cogan, L.; Scanaill, C.; Kenny, R. Quantitative falls risk assessment using the timed up and go test. *IEEE Trans. Biomed. Eng.* **2010**, *57*, 2918–2926. [CrossRef]
46. Mickle, K.; Munro, B.; Lord, S.; Menz, H.; Steele, J. Foot pain, plantar pressures, and falls in older people: A prospective study. *J. Am. Geriatr. Soc.* **2010**, *58*, 1936–1940. [CrossRef]
47. Kirkwood, R.; Trede, R.; Moreira, B.; Kirkwood, S.; Pereira, L. Decreased gastrocnemius temporal muscle activation during gait in elderly women with history of recurrent falls. *Gait Posture* **2011**, *34*, 60–64. [CrossRef]
48. Lázaro, M.; González, A.; Latorre, G.; Fernández, C.; Ribera, J. Postural stability in the elderly: Fallers versus non-fallers. *Eur. Geriatr. Med.* **2011**, *2*, 1–5. [CrossRef]
49. Lugade, V.; Lin, V.; Chou, L. Center of mass and base of support interaction during gait. *Gait Posture* **2011**, *33*, 406–411. [CrossRef]
50. Panzer, V.; Wakefield, D.; Hall, C.; Wolfson, L. Mobility assessment: Sensitivity and specificity of measurement sets in older adults. *Arch. Phys. Med. Rehabil.* **2011**, *92*, 905–912. [CrossRef] [PubMed]
51. Scanaill, C.; Greene, B.; Doheny, E.; O'Donovan, K.; O'Shea, T.; O´Donovan, A.; Foran, T.; Cunningham, C.; Kenny, R. Clinical gait assessment of older adults using open platform tools. In Proceedings of the 2011 Annual International Conference of the IEEE Engineering in Medicine and Biology Society, Boston, MA, USA, 30 August–3 September 2011; IEEE: Piscataway, NJ, USA, 2011; pp. 462–465. [CrossRef]
52. Uemura, K.; Yamada, M.; Nagai, K.; Tanaka, B.; Mori, S.; Ichihashi, N. Fear of falling is associated with prolonged anticipatory postural adjustment during gait initiation under dual-task conditions in older adults. *Gait Posture* **2012**, *35*, 282–286. [CrossRef] [PubMed]
53. Chen, T.; Chou, L. Altered center of mass control during sit-to-walk in elderly adults with and without a history of falling. *Gait Posture* **2013**, *38*, 696–701. [CrossRef] [PubMed]
54. Chen, T.; Chang, C.; Chou, L. Sagittal plane center of mass movement strategy and joint kinetics during sit-to-walk in elderly fallers. *Clin. Biomech.* **2013**, *28*, 807–812. [CrossRef] [PubMed]
55. Chiu, S.; Chou, L. Variability in inter-joint coordination during walking of elderly adults and its association with clinical balance measures. *Clin. Biomech.* **2013**, *28*, 454–458. [CrossRef] [PubMed]
56. Fritz, N.; Worstell, A.; Kloos, A.; Siles, A.; White, S.; Kegelmeyer, D. Backward walking measures are sensitive to age-related changes in mobility and balance. *Gait Posture* **2013**, *37*, 593–597. [CrossRef] [PubMed]
57. Weiss, A.; Brozgol, M.; Dorfman, M.; Herman, T.; Shema, S.; Giladi, N.; Hausdorff, J. Does the evaluation of gait quality during daily life provide insight into fall risk? a novel approach using 3-day accelerometer recordings. *Neurorehabil. Neural Repair* **2013**, *27*, 742–752. [CrossRef] [PubMed]
58. Marques, N.; Hallal, C.; Crozara, L.; Morcelli, M.; Karuka, A.; Navega, M.; Gonçalves, M. Lower limb strength is associated with gait biomechanical abnormalities in older female fallers and non-fallers. *Isokinet. Exerc. Sci.* **2013**, *21*, 151–159. [CrossRef]
59. Marques, N.; LaRoche, D.; Hallal, C.; Crozara, L.; Morcelli, M.; Karuka, A.; Navega, M.; Gonçalves, M. Association between energy cost of walking, muscle activation, and biomechanical parameters in older female fallers and non-fallers. *Clin. Biomech.* **2013**, *28*, 330–336. [CrossRef] [PubMed]
60. Ayoubi, F.; Launay, C.; Kabeshova, A.; Fantino, B.; Annweiler, C.; Beauchet, O. The influence of fear of falling on gait variability: Results from a large elderly population-based cross-sectional study. *J. Neuroeng. Rehabil.* **2014**, *11*, 128. [CrossRef] [PubMed]
61. Barelle, C.; Houel, N.; Koutsouris, D. A cluster analysis approach for the determination of a fall risk level classification. In Proceedings of the IEEE 19th International Workshop on Computer Aided Modeling and Design of Communication Links and Networks (CAMAD), Athens, Greece, 1–3 December 2014; IEEE: Piscataway, NJ, USA, 2014; pp. 130–134. [CrossRef]

62. Iwata, A.; Sano, Y.; Inoue, J.; Iwata, H.; Fuchioka, S. Movement velocity is a sensitive risk factor of falls in high-functioning older adults. *J. Am. Geriatr. Soc.* **2014**, *62*, 1595–1596. [CrossRef] [PubMed]
63. Kobayashi, Y.; Hobara, H.; Matsushita, S.; Mochimaru, M. Key joint kinematic characteristics of the gait of fallers identified by principal component analysis. *J. Biomech.* **2014**, *47*, 2424–2429. [CrossRef]
64. König, N.; Taylor, W.; Armbrecht, G.; Dietzel, R.; Singh, N. Identification of functional parameters for the classification of older female fallers and prediction of 'first-time' fallers. *J. R. Soc. Interface* **2014**, *11*, 20140353. [CrossRef]
65. Mignardot, J.; Deschamps, T.; Barrey, E.; Auvinet, B.; Gilles Berrut, G.; Cornu, C.; Constans, T.; Decker, L. Gait disturbances as specific predictive markers of the first fall onset in elderly people: A two-year prospective observational study. *Front. Aging Neurosci.* **2014**, *6*, 22. [CrossRef]
66. Cebolla, E.; Rodacki, A.; Bento, P. Balance, gait, functionality and strength: Comparison between elderly fallers and non-fallers. *Braz. J. Phys. Ther.* **2015**, *19*, 146–151. [CrossRef]
67. MacAulay, R.; Allaire, T.; Brouillette, R.; Foil, H.; Bruce-Keller, A.; Han, H.; Johnson, W.; Keller, J. Longitudinal assessment of neuropsychological and temporal/spatial gait characteristics of elderly fallers: Taking it all in stride. *Front. Aging Neurosci.* 2015. [CrossRef]
68. Rispens, S.; Schooten, K.; Pijnappels, M.; Daffertshofer, A.; Beek, P.; Dieën, J. Identification of fall risk predictors in daily life measurements. *Neurorehabil. Neural Repair* **2015**, *29*, 54–61. [CrossRef]
69. Wright, R.; Peters, D.; Robinson, P.; Watt, T.; Hollands, M. Older adults who have previously fallen due to a trip walk differently than those who have fallen due to a slip. *Gait Posture* **2015**, *41*, 164–169. [CrossRef]
70. Bounyong, S.; Adachi, S.; Ozawa, J.; Yamada, Y.; Kimura, M.; Watanabe, Y.; Yokoyama, K. Fall risk estimation based on co-contraction of lower limb during walking. In Proceedings of the IEEE International Conference on Consumer Electronics (ICCE), Las Vegas, NV, USA, 7–11 January 2016; IEEE: Piscataway, NJ, USA, 2016; pp. 331–332. [CrossRef]
71. Fujimoto, M.; Chou, L. Sagittal plane momentum control during walking in elderly fallers. *Gait Posture* **2016**, *45*, 121–126. [CrossRef]
72. Ihlen, E.; Weiss, A.; Bourke, A.; Helbostad, J.; Hausdorff, J. The complexity of daily life walking in older adult community-dwelling fallers and non-fallers. *J. Biomech.* **2016**, *49*, 1420–1428. [CrossRef]
73. Howcroft, J.; Kofman, J.; Lemaire, E.; McIlroy, W. Analysis of dual-task elderly gait in fallers and non-fallers using wearable sensors. *J. Biomech.* **2016**, *49*, 992–1001. [CrossRef]
74. Rinaldi, N.; Moraes, R. Older adults with a history of falls are unable to perform walking and prehension movements simultaneously. *Neuroscience* **2016**, *316*, 249–260. [CrossRef]
75. Rinaldi, N.; Emmerik, R.; Moraes, R. Changes in interlimb coordination during walking and grasping task in older adult fallers and non-fallers. *Hum. Mov. Sci.* **2017**, *55*, 121–137. [CrossRef]
76. Bizovskaa, L.; Svoboda, Z.; Vuillerme, N.; Janura, M. Multiscale and shannon entropies during gait as fall risk predictors—A prospective study. *Gait Posture* **2017**, *52*, 5–10. [CrossRef] [PubMed]
77. Worms, J.; Stins, J.; Wegen, E.; Loram, I.; Beek, P. Influence of focus of attention, reinvestment, and fall history on elderly gait stability. *Physiol. Rep.* **2017**, *5*, e13061. [CrossRef] [PubMed]
78. Worms, J.; Stins, J.; Wegen, E.; Verschueren, S.; Beek, P.; Loram, I. Effects of attentional focus on walking stability in the elderly. *Gait Posture* **2017**, *55*, 94–99. [CrossRef]
79. Marques, N.; Hallal, C.; Spinoso, D.; Morcelli, M.; Crozara, L.; Gonçalves, M. Applying different mathematical variability methods to identify older fallers and non-fallers using gait variability data. *Aging Clin. Exp. Res.* **2017**, *29*, 473–481. [CrossRef] [PubMed]
80. Júnior, R.; Porto, J.; Marques, N.; Magnani, P.; Abreu, D. The effects of a simultaneous cognitive or motor task on the kinematics of walking in older fallers and non fallers. *Hum. Mov. Sci* **2017**, *51*, 146–152. [CrossRef] [PubMed]
81. Svoboda, Z.; Bizovska, L.; Janura, M.; Kubonova, E.; Janurova, K.; Vuillerme, N. Variability of spatial-temporal gait parameters and center of pressure displacements during gait in elderly fallers and non-fallers: A 6-month prospective study. *PLoS ONE* **2017**, *12*, e0171997. [CrossRef] [PubMed]
82. Allen, J.; Franz, J. The motor repertoire of older adult fallers may constrain their response to balance perturbations. *J. Neurophysiol.* **2018**, *120*, 2368–2378. [CrossRef]
83. Benson, L.; Cobb, S.; Hyngstrom, A.; Keenan, K.; Luo, J.; O'Connor, K. Identifying trippers and non-trippers based on knee kinematics during obstacle-free walking. *Hum. Mov. Sci.* **2018**, *62*, 58–66. [CrossRef]
84. Kwon, M.; Kwon, Y.; Park, Y.; Kim, J. Comparison of gait patterns in elderly fallers and non-fallers. *Technol. Health Care* **2018**, *26*, 427–436. [CrossRef] [PubMed]
85. Marques, N.; Spinoso, D.; Cardoso, B.; Moreno, V.; Kuroda, M.; Navega, M. Is it possible to predict falls in older adults using gait kinematics? *Clin. Biomech.* **2018**, *59*, 15–18. [CrossRef]
86. Howcroft, J.; Lemaire, E.; Kofman, J.; McIlroy, W. Dual-task elderly gait of prospective fallers and non-fallers: A wearable-sensor based analysis. *Sensors* **2018**, *18*, 1275. [CrossRef] [PubMed]
87. Thompson, J.; Plummer, P.; Franz, J. Age and falls history effects on antagonist leg muscle coactivation during walking with balance perturbations. *Clin. Biomech.* **2018**, *59*, 94–100. [CrossRef]
88. Qiao, M.; Truong, K.; Franz, J. Does local dynamic stability during unperturbed walking predict the response to balance perturbations? an examination across age and falls history. *Gait Posture* **2018**, *62*, 80–85. [CrossRef]

89. Watanabe, K. Relationship between toe clearance strategy and regional regulation of rectus femoris muscle during swing phase in prolonged walking in young and older adults. *Front. Physiol* **2018**, *9*, 1274. [CrossRef]
90. Bueno, G.; Ribeiro, D.; Gervásio, F.; Correia, A.; Menezes, R. Gait profile score identifies changes in gait kinematics in nonfaller, faller, and recurrent faller older adults women. *Gait Posture* **2019**, *72*, 76–81. [CrossRef]
91. Gillain, S.; Boutaayamou, M.; Schwartz, C.; Dardenne, N.; Bruyère, O.; Brüls, O.; Croisier, J.; Salmon, E.; Reginster, J.; Garraux, G.; et al. Gait symmetry in the dual-task condition as a predictor of future falls among independent older adults: A 2-year longitudinal study. *Aging Clin. Exp. Res.* **2019**, *31*, 1057–1067. [CrossRef]
92. Mak, T.; Young, W.; Lam, W.; Tse, A.; Wong, T. The role of attentional focus on walking efficiency among older fallers and non-fallers. *Age Ageing* **2019**, *48*, 811–816. [CrossRef]
93. Yamagata, M.; Tateuchi, H.; Shimizu, I.; Saeki, J.; Ichihashi, N. The relation between limb segment coordination during walking and fall history in community-dwelling older adults. *J. Biomech.* **2019**, *93*, 94–100. [CrossRef]
94. Yamagata, M.; Tateuchi, H.; Shimizu, I.; Ichihashi, N. The effects of fall history on kinematic synergy during walking. *J. Biomech.* **2019**, *82*, 204–210. [CrossRef]
95. Gonzalez, M.; Gates, D.; Rosenblatt, N. The impact of obesity on gait stability in older adults. *J. Biomech.* **2020**, *100*, 109585. [CrossRef]
96. Yamagata, M.; Tateuchi, H.; Shimizu, I.; Saeki, J.; Ichihashi, N. The relation between kinematic synergy to stabilize the center of mass during walking and future fall risks: A 1-year longitudinal study. *BMC Geriatr.* **2021**, *21*, 240. [CrossRef]
97. Pol, F.; Forghany, S.; Hosseini, S.; Taheri, A.; Menz, H. Structural and functional foot and ankle characteristics associated with falls in older people. *Gait Posture* **2021**, *88*, 78–83. [CrossRef]
98. Sadeghi, H.; Shojaedin, S.; Abbasi, A.; Alijanpour, E.; Vieira, M.; Svoboda, Z.; Nazarpour, K. Lower-extremity intra-joint coordination and its variability between fallers and non-fallers during gait. *Appl. Sci.* **2021**, *11*, 2840. [CrossRef]
99. Figueiredo, A.; Balbinot, G.; Brauner, F.; Schiavo, A.; Urbanetto, M.; Mestriner, R. History of falls alters movement smoothness and time taken to complete a functional mobility task in the oldest-old: A case-control study. *Exp. Gerontol.* **2022**, *167*, 111918. [CrossRef]
100. Nascimento, M.; Gouveia, E.; Gouveia, B.; Marques, A.; Martins, F.; Przednowek, K.; França, C.; Peralta, M.; Ihle, A. Associations of gait speed, cadence, gait stability ratio, and body balance with falls in older adults. *Int. J. Environ. Res. Public Health* **2022**, *19*, 13926. [CrossRef]
101. Yoshida, K.; Commandeur, D.; Hundza, S.; Klimstra, M. Detecting differences in gait initiation between older adult fallers and non-fallers through multivariate functional principal component analysis. *J. Biomech.* **2022**, *144*, 111342. [CrossRef]
102. Baba, T.; Watanabe, M.; Ogihara, H.; Handa, S.; Sasamoto, K.; Okada, S.; Okuizumi, H.; Teiji Kimura, T. Validity of temporo-spatial characteristics of gait as an index for fall risk screening in community-dwelling older people. *J. Phys. Ther. Sci.* **2023**, *35*, 2022–2153. [CrossRef]
103. Pijnappels, M.; Bobbert, M.; Dieën, J. Push-off reactions in recovery after tripping discriminate young subjects, older non-fallers and older fallers. *Gait Posture* **2005**, *21*, 388–394. [CrossRef] [PubMed]
104. Bhatt, T.; Espy, D.; Yang, F.; Pai, Y. Dynamic gait stability, clinical correlates, and prognosis of falls among community-dwelling older adults. *Arch. Phys. Med. Rehabil.* **2011**, *92*, 799–805. [CrossRef] [PubMed]
105. Pavol, M.; Owings, M.; Foley, K.; Grabiner, M. Gait characteristics as risk factors for falling from trips induced in older adults. *J. Gerontol. A Biol. Sci. Med. Sci.* **1999**, *54*, M583–M590. [CrossRef] [PubMed]
106. Lockhart, T.; Woldstad, J.; Smith, J. Effects of age-related gait changes on the biomechanics of slips and falls. *Ergonomics* **2003**, *46*, 1136–1160. [CrossRef] [PubMed]
107. Pavol, M.; Owings, M.; Foley, K.; Grabiner, M. Mechanisms leading to a fall from an induced trip in healthy older adults. *J. Gerontol. A Biol. Sci. Med. Sci.* **2001**, *56*, M428–M437. [CrossRef] [PubMed]
108. Espy, D.; Pai, Y. Relationship of frontal plane center of mass control variables to slip outcomes. *J. Geriatr. Phys. Ther.* **2007**, *30*, 137. [CrossRef]
109. Yang, F.; Pai, Y. Can stability really predict an impending slip-related fall among older adults? *J. Biomech.* **2014**, *47*, 3876–3881. [CrossRef] [PubMed]
110. Sawers, A.; Pai, Y.; Bhatt, T.; Ting, L. Neuromuscular responses differ between slip-induced falls and recoveries in older adults. *J. Neurophysiol.* **2017**, *117*, 509–522. [CrossRef]
111. Sawers, A.; Bhatt, T. Neuromuscular determinants of slip-induced falls and recoveries in older adults. *J. Neurophysiol.* **2018**, *120*, 1534–1546. [CrossRef]
112. Bruijn, S.; Sloot, L.; Kingma, I.; Pijnappels, M. Contribution of arm movements to balance recovery after tripping in older adults. *J. Biomech.* **2022**, *133*, 110981. [CrossRef] [PubMed]
113. Wang, S.; Miranda, F.; Wang, Y.; Rasheed, R.; Bhatt, T. Near-fall detection in unexpected slips during over-ground locomotion with body-worn sensors among older adults. *Sensors* **2022**, *22*, 3334. [CrossRef] [PubMed]
114. Wang, S.; Bhatt, T. Kinematic measures for recovery strategy identification following an obstacle-induced trip in gait. *J. Mot. Behav.* **2023**, *55*, 193–201. [CrossRef] [PubMed]
115. Fukuchi, C.; Fukuchi, R.; Duarte, M. Effects of walking speed on gait biomechanics in healthy participants: A systematic review and meta-analysis. *Syst. Rev.* **2019**, *8*, 153. [CrossRef] [PubMed]

116. Moosabhoy, M.; Gard, S. Methodology for determining the sensitivity of swing leg toe clearance and leg length to swing leg joint angles during gait. *Gait Posture* **2006**, *24*, 493–501. [CrossRef] [PubMed]
117. Aleixo, P.; Atalaia, T.; Patto, J.; Abrantes, J. Minimum toe clearance lower limb angle joints strategies within rheumatoid arthritis post-menopausal women. *Gait Posture* **2019**, *73*, 476–477. [CrossRef]
118. Bhudarally, M.; Aleixo, P.; Atalaia, T.; Abrantes, J. Swing limb biomechanics and minimum toe clearance in rheumatoid arthritis post-menopausal women fallers and non-fallers. *Gait Posture* **2020**, *81*, 40–41. [CrossRef]
119. Aleixo, P.; Atalaia, T.; Patto, J.; Abrantes, J. The effect of a proprioceptive exercises programme on disease activity and gait biomechanical parameters of post-menopausal women with rheumatoid arthritis. In *Rheumatoid Arthritis*; Toumi, H., Ed.; IntechOpen: London, UK, 2022. [CrossRef]
120. Aleixo, P.; Abrantes, J. Proprioceptive and strength exercise guidelines to prevent falls in the elderly related to biomechanical movement characteristics. *Healthcare* **2024**, *12*, 186. [CrossRef]
121. Aleixo, P.; Atalaia, T.; Abrantes, J. Dynamic joint stiffness: A critical review. In *Advances in Medicine and Biology*; Berhardt, V., Ed.; Nova Science Publishers, Inc.: Hauppauge, NY, USA, 2021; Volume 175, pp. 1–96.
122. Aleixo, P.; Atalaia, T.; Patto, J.; Abrantes, J. Sit-to-stand of rheumatoid arthritis post-menopausal women vs. healthy post-menopausal—Antero-posterior displacement of centre of gravity and centre of pressure. *Gait Posture* **2018**, *65*, 384–385. [CrossRef]
123. Caron, O.; Gelat, T.; Rougier, P.; Blanchi, J. A comparative analysis of the center of gravity and center of pressure trajectory path lengths in standing posture: An estimation of active stiffness. *J. Appl. Biomech.* **2000**, *16*, 234–247. [CrossRef] [PubMed]
124. Takeda, K.; Mani, H.; Hasegawa, N.; Sato, Y.; Tanaka, S.; Maejima, H.; Asaka, T. Adaptation effects in static postural control by providing simultaneous visual feedback of center of pressure and center of gravity. *J. Physiol. Anthr.* **2017**, *36*, 31. [CrossRef] [PubMed]
125. Schmitt, A.; Baudendistel, S.; Lipat, A.; White, T.; Raffegeau, T.; Hass, C. Walking indoors, outdoors, and on a treadmill: Gait differences in healthy young and older adults. *Gait Posture* **2021**, *90*, 468–474. [CrossRef] [PubMed]
126. Di Paolo, S.; Lopomo, N.; Villa, F. Rehabilitation and return to sport assessment after anterior cruciate ligament injury: Quantifying joint kinematics during complex high-speed tasks through wearable sensors. *Sensors* **2021**, *21*, 2331. [CrossRef] [PubMed]
127. Müller, B.; Ilg, W.; Giese, M.; Ludolph, N. Validation of enhanced kinect sensor based motion capturing for gait assessment. *PLoS ONE* **2017**, *12*, e0175813. [CrossRef]
128. Springer, S.; Seligmann, G. Validity of the kinect for gait assessment: A focused review. *Sensors* **2016**, *16*, 194. [CrossRef]
129. Aleixo, P.; Patto, J.; Abrantes, J. Dynamic joint stiffness of the ankle in rheumatoid arthritis postmenopausal women fallers and non-fallers. *Gait Posture* **2017**, *57*, 324–325. [CrossRef]

biomechanics

MDPI

Article

People with Parkinson's Disease Are Able to Couple Eye Movements and Postural Sway to Improve Stability

Fabio Augusto Barbieri [1,*], Paula Favaro Polastri [2], José Angelo Barela [3], Cédrick T. Bonnet [4], Matheus Belizario Brito [2] and Sergio Tosi Rodrigues [2]

1 Human Movement Research Laboratory (MOVI-LAB), Department of Physical Education, São Paulo State University (UNESP), Bauru 17033-360, SP, Brazil
2 Laboratory of Information, Vision and Action (LIVIA), Department of Physical Education, São Paulo State University (UNESP), Bauru 17033-360, SP, Brazil; paula.polastri@unesp.br (P.F.P.); matheus.belizario@unesp.br (M.B.B.); sergio.tosi@unesp.br (S.T.R.)
3 Institute of Biosciences, Department of Physical Education, São Paulo State University (UNESP), Rio Claro 13506-900, SP, Brazil; jose.barela@unesp.br
4 UMR 9193-SCALab-Sciences Cognitives et Sciences Affectives, National Center for the Scientific Research (CNRS), Univ. Lille, F-59000 Lille, France; cedrick.bonnet@univ-lille.fr
* Correspondence: fabio.barbieri@unesp.br

Abstract: Considering that people with Parkinson's disease (PD) experience challenges in the control of both balance and eye movements, this study investigated the effects of saccadic eye movements on body sway in people with PD in two bases of support positions (side-by-side and tandem stances). Ten people with PD and 11 healthy individuals performed (a) fixation; (b) horizontal saccadic eye movements to the right and left; and (c) vertical saccadic eye movements up and down. The protocol for each postural task consisted of one block of six trials, making a total of 12 trials. Body sway and gaze parameters were measured during the trials. In both people with PD and healthy individuals, anterior–posterior body sway was significantly reduced in horizontal saccadic eye movements in contrast to fixation, regardless of the body position (side-by-side and tandem stances). Furthermore, vertical saccadic eye movements increased the area of sway in contrast to horizontal ones (and not to fixation) in people with PD. In addition, people with PD showed a higher number of fixations in all experimental conditions, without changes in the mean duration of fixations in both body positions. In conclusion, individuals with PD can improve body sway by coupling eye movements and postural sway when performing horizontal saccadic eye movements but not when performing vertical saccadic eye movements.

Keywords: posture; Parkinson's disease; gaze; saccadic eye movements; body sway

1. Introduction

The coordination between postural control and visual system is essential for successfully performing multiple tasks and achieving environmental goals [1]. Young adults are able to control body sway to facilitate eye movements during visual search tasks [1] and to use reliable visual information to enhance body sway control [2]. People with Parkinson's disease (PD) experience challenges in the control of both balance [3] and eye movements [4].

People with PD display larger body sway magnitude [5–7], higher body sway acceleration and velocity, and lower complexity and adaptability [8] in postural control function while standing upright compared to age-matched neurologically healthy controls [9]. In addition, their limits of stability are reduced [10] due to a disruption of the precisely coordinated execution of agonist and antagonist muscles [11]. These postural impairments are problematic because they can lead to falls [12], which are related to asymmetric control of posture in people with PD [13]. The postural sway of people with PD is higher during challenging upright postural tasks, such as tandem position [14,15]. Poor postural control performance in people with PD is not surprising, considering the many changes in sensory systems [16].

It is known that individuals with PD particularly rely on visual information to compensate for deficits in postural sway control [17]. However, eye movement abnormalities in people with PD are extensively discussed in the literature, including a decreased range of eye movements and prolonged saccadic latency [4,18]. In addition, this population shows hypometria during the execution of voluntary saccades (remembered, predictive, and antisaccades) [19] and impairment of antisaccade latencies, which is an indirect marker of impaired anticipatory postural adjustments [20]. The deficits in voluntary saccade movements in PD are attributed to pathological involvement at the brainstem and basal ganglia levels [20]. While the basal ganglia mediate saccade amplitude and latency [18], the cerebellum is involved in saccade accuracy [21]. However, it is not clear yet whether and how aspects of saccade and postural control are conjointly impaired in PD. A better understanding of how eye movements and postural sway are linked in PD should help to develop more effective strategies to minimize balance impairments in this population.

In both young adults [22,23] and healthy older adults [24,25], it has been shown that body sway is reduced when performing saccadic eye movements. This reduction in postural sway allows the brain to shift gaze accurately, indicating a functional connection between posture and gaze control [1]. However, the impact of saccadic eye movements on body sway depends on the direction of the saccadic eye movements (horizontal vs. vertical) and the level of difficulty of the standing posture in both young and older adults. Vertical saccadic eye movements are more challenging for the eye movement system than horizontal saccade ones, leading to a delayed gaze response [26]. Greater postural instability seems to impair eye movements in the vertical direction, particularly in older adults [25]. When exposed to a challenging standing position (e.g., tandem stance), older adults exhibited greater variability in gaze during vertical saccadic eye movements, which is associated with larger head movements compared to young adults [25]. Furthermore, older adults experience an increase in the effects of saccadic eye movements on body sway when performing a more challenging postural task, adopting a more rigid postural control strategy [24].

This stabilization of postural control could be attributed to both afferent and efferent mechanisms for eye movements. In fact, the afferent motion perception mechanism occurs when the central nervous system uses optic flow information to minimize retinal slip and stabilize the distance between the eye and the visual scene; while the efferent motion perception mechanism occurs when the central nervous system uses the copy of motor commands (i.e., "efference copy") or extraocular muscle afferents that follow eye movements to stabilize posture [2]. Another explanation for the reduction in body sway during saccadic eye movements in healthy adults may be associated with the achievement of a goal in a supra-postural task [27]. This type of activity includes tasks or behavioral goals that are subordinated to the control of posture [27]. In this case, postural coordination and supra-postural performance have a hierarchical relationship, where posture coordination serves as a means to succeed in supra-postural tasks. The main viewpoint is that the neurologically healthy individual improves postural stability to facilitate, at least in part, the performance of this type of task [28]. Both explanations consider additional resources available, such as sensory cues, attentional efforts, and cognitive engagement, as essential for decreasing postural sway and favoring the spatial accuracy of the saccade in terms of target location [23,29].

Considering that postural sway is decreased in performing saccadic eye movements in healthy young and older adults, testing the impact of such saccadic eye movements is important in people with PD who often experience larger body sway, especially in challenging stances (i.e., semi-tandem position) [14,15], and deficits in voluntary saccade movements [4,18], which increase the risk of falls [12]. In this study, we investigated the effects of horizontal and vertical saccadic eye movements on body sway in people with PD (under dopaminergic medication) and neurologically healthy individuals (control group) during quiet stance tasks in two bases of support positions (side-by-side and tandem stances).

We hypothesized that (i) the brain would decrease body sway to perform horizontal saccadic eye movement compared to the eye fixation condition in neurologically healthy individuals, with a greater reduction during tandem stance than side-by-side stance [24], but we did not expect changes in body sway in people with PD during horizontal saccadic eye movement due to their postural [14,15] and visual deficits [4,18]; (ii) the brain would not reduce body sway during vertical saccadic eye movement compared to the eye fixation condition in both neurologically healthy individuals and people with PD because of the weaker functional connection between posture and gaze control in this scenario [25]; (iii) people with PD would show poorer gaze performance, with an increased number of fixations and decreased mean duration of fixations during both horizontal and vertical saccadic eye movements compared to neurologically healthy individuals, with a greater impact during tandem stance than side-by-side stance; and iv) both groups would show worse gaze performance during vertical saccadic eye movements compared to horizontal saccadic eye movements, considering that the first type of eye movement is more challenging for the eye movement system [26].

2. Materials and Methods

2.1. Participants

Ten individuals diagnosed with PD (PD group) (8 men/2 women) and 11 neurologically healthy individuals (7 men/4 women) participated in this study. The number of participants was determined using a power analysis that used an alpha level of 0.05, effect size of 0.93 and a power of 90% (G-power®). The analysis was based on the mean velocity of the center of pressure (CoP) from Polastri and colleagues' study [25], which determined that a minimum of nine participants in each group was needed for the study.

All participants included in this study were over the age of 60 and were able to stand independently during the postural task. The PD group included only those who had received a confirmed diagnosis of idiopathic PD from a specialist according to the criteria of the UK Parkinson's Disease Brain Bank Field [30], classified between stages 1 and 3 on the Hoehn and Yahr (H&Y) scale [31] and treated with dopaminergic medication. Participants with cognitive deficits (defined as a score below 24 on the Mini-Mental State Examination [32]) (for both groups), rheumatic or orthopedic diseases that impaired the performance of the postural task (for both groups), or any neurological diseases (for the control group) or other neurological diseases (for the PD group) were excluded from the study. To ensure consistent visual acuity among participants, we conducted the Snellen test and only selected individuals with visual acuity scores between 20/20 and 20/30 to participate in the study. Five individuals with PD and three neurological healthy individuals did not meet the inclusion/exclusion criteria and were therefore excluded from the study.

All participants signed an informed consent form and heard a thorough explanation of the research procedures before signing the document. The study received approval from the University Ethics and Research Committee (#11322/46/01/12).

2.2. Clinical Evaluation

A specialist in movement disorders conducted a cognitive evaluation of both the control group and the PD group. The control group was evaluated using the Mini-Mental State Examination and an anamnesis. The PD group was evaluated using a procedure similar to that for the control group, in addition to the H&Y scale and the Unified Parkinson's Disease Rating Scale—UPDRS-III [33]. The PD group was assessed under dopaminergic medication ("ON" state) [34].

2.3. Postural Task and Eye Movements

During the experiment, each participant was asked to stand barefoot on a force plate (AMTI-AccuGait). An LCD monitor (37.5 cm × 30 cm, LG, Faltron L1952H, 50/60 Hz, 0.8 A) was placed at eye level, and a red dot with a 2 cm diameter was displayed on the monitor, which was 1 m away from the participant's eyes. The target was presented on a white background, and its subtended visual angle was 1.15° (relative to looking straight ahead). The stimulus was generated using Flash Mx 7.2 software (Macromedia, Portland, OR, USA).

The participants completed the trials of the two postural tasks in a sequential order, and the three visual conditions in a randomized order. The protocol for each postural task consisted of one block of six trials, making a total of 12 trials. After every three trials, the participant was given a one-minute rest period to avoid fatigue or tiredness that could potentially deteriorate their performance. Each trial lasted 70 s.

The participants performed two types of postural tasks under eye movement conditions: (1) to stand with the feet in a side-by-side stance, with the feet parallel and aligned with the shoulders; (2) to stand with the feet in a tandem stance, with the least affected or dominant foot aligned in front of the most affected or non-dominant foot (Figure 1d,e). The three visual tasks were (a) fixation, where the participant fixed their gaze on a single target displayed at the center of the screen (Figure 1b); (b) horizontal saccadic eye movements to the right and left, with the subject tracking the targets in those directions on the screen (Figure 1a); (c) vertical saccadic eye movements up and down, with the individual keeping their eye on the target positioned above or below the screen (Figure 1c). During horizontal and vertical saccadic eye movements, (i) the angle between targets was 11° to avoid head movements, and (ii) the target appeared first on the left side of the monitor, 9.75 cm away from the center, and then disappeared and reappeared immediately on the opposite side (i.e., the right side), also 9.75 cm away from the center.

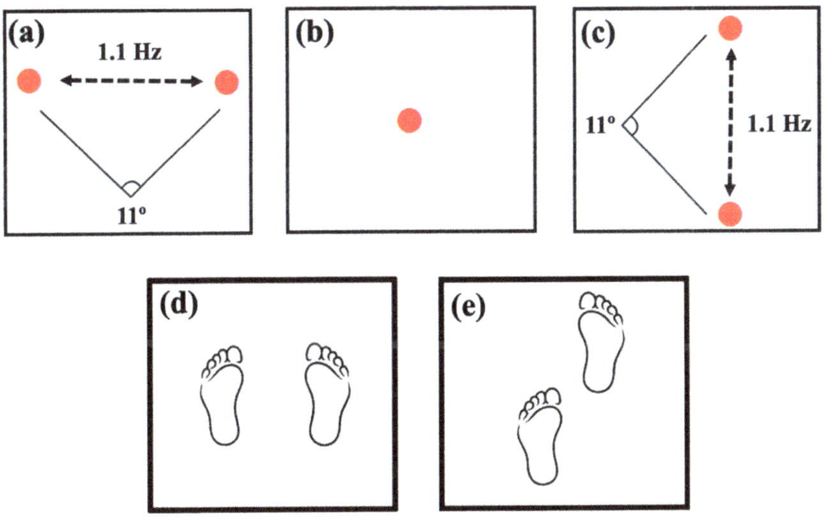

Figure 1. A representation of postural tasks and visual conditions: (**a**) horizontal saccadic eye movements, (**b**) fixation, (**c**) vertical saccadic eye movements, (**d**) feet in side-by-side stance, and (**e**) tandem stance. The red dot is the target for eye movement tasks.

To determine the lower-limb preference of the control group, the participants were asked to kick a ball. The limb used to kick the ball was considered the preferred limb [35]. For the people with PD, the most affected limb was determined using items 20–23 and 25–26 of UPDRS-III. The value of the right limb was subtracted from the value of the left limb in each item. If the result of this calculation was positive or negative, the most affected limb was the right or left limb, respectively [35].

2.4. Data Analysis

The force plate measures the forces (Fx, Fy and Fz) and moment components (Mx, My and Mz) to calculate the displacement of the CoP in the anterior–posterior (AP) and medial–lateral (ML) axes. The CoP displacement was filtered using a second low-pass Butterworth filter with a 5 Hz cutoff frequency and a fourth-order zero-lag. A head-mounted eye tracking system (model H6, Applied Science Laboratory, Billerica, MA, USA) was used to measure eye movements with a precision of 1° of visual angle. The system calibration was performed from the fixation of nine points displayed in a 3 by 3 grid and checked in each trial. The sampling frequency was 60 Hz. Although each trial lasted 70 s, the first 10 s were not considered in the analyses.

The following CoP variables were calculated for both AP and ML directions: displacement of sway—the total path length of the CoP along the support base in each direction; mean velocity of sway—the division of the total sway in each direction by the duration of the trial; and root mean square (RMS)—the mean variability of the displacement along the trial. In addition, the area was calculated as 95% of the ellipse area that the CoP covered. To analyze the data from the force platform, a group of specific programs written in MATLAB R_2022a (MathWorks, Inc®, Natick, MA, USA) was used.

Eye movement analysis was conducted using Applied Sciences Laboratories Results Plus software® (Billerica, MA, USA) and MATLAB (MathWorks, Inc®). The following gaze parameters were calculated for both conditions: number of fixations (the total number of fixations during the trials), and mean duration of fixations (the average of all fixation durations in each trial). To calculate these gaze parameters, an eye fixation occurred when the value of two times point of gaze standard deviation was less than one degree of visual angle in the horizontal axis and one degree of visual angle in the vertical axis over 100 ms.

2.5. Statistical Analysis

CoP and gaze variables were pre-tested with the Shapiro–Wilk test to verify the normality of the data and with Levene's test to verify the equality of variance. Independent t-tests were performed to compare anthropometric and cognitive status between the PD group and the control group. Three-way ANOVAs were conducted to test differences between groups, and within-group conditions, for each CoP- and gaze-dependent variable. The study examined the effects of groups (PD vs. control), visual tasks (fixation, horizontal, and vertical saccadic eye movements) and base of support (side-by-side vs. tandem). The last two factors were measured repeatedly. When the ANOVA showed significant differences between variables, post hoc tests with Bonferroni adjustments were performed. For gaze-dependent variables, visual tasks had only two levels (horizontal and vertical saccadic eye movements). The effect size (η^2, partial eta-squared) was also calculated for each statistical analysis and was interpreted as a small effect if it was <0.06, moderate if it was within >0.06 and <0.14, and large if it was >0.14 [36]. The significance level was $p < 0.05$ (SPSS, version 26.0).

3. Results

Table 1 depicts the anthropometric and clinical status of the PD group and control group. There were no significant differences in age, body mass, height and cognitive status between the two groups (all $p > 0.05$).

Table 1. Anthropometric and clinical characteristics of the participants.

	PD Group	Control Group
Age (years)	66.0 ± 6.0	69.8 ± 3.4
Body mass (kg)	67.4 ± 16.3	-

Table 1. *Cont.*

	PD Group	Control Group
Height (m)	1.64 ± 0.05	-
H&Y stage	2.2 ± 0.4	-
UPDRS-III (pts)	19.3 ± 6.1	-
MMSE (pts)	28.2 ± 1.8	28.5 ± 1.1

H&Y—Hoehn and Yahr scale; Unified Parkinson's Disease Rating Scale—motor part—UPDRS-III; MMSE—Mini-Mental State Examination.

3.1. CoP Parameters

Figures 2 and 3 depict the CoP variables for side-by-side and tandem support positions during fixation and eye movements in horizontal and vertical directions. ANOVAs did not reveal a significant three-factor interaction for the group and base of support and visual task ($p > 0.05$).

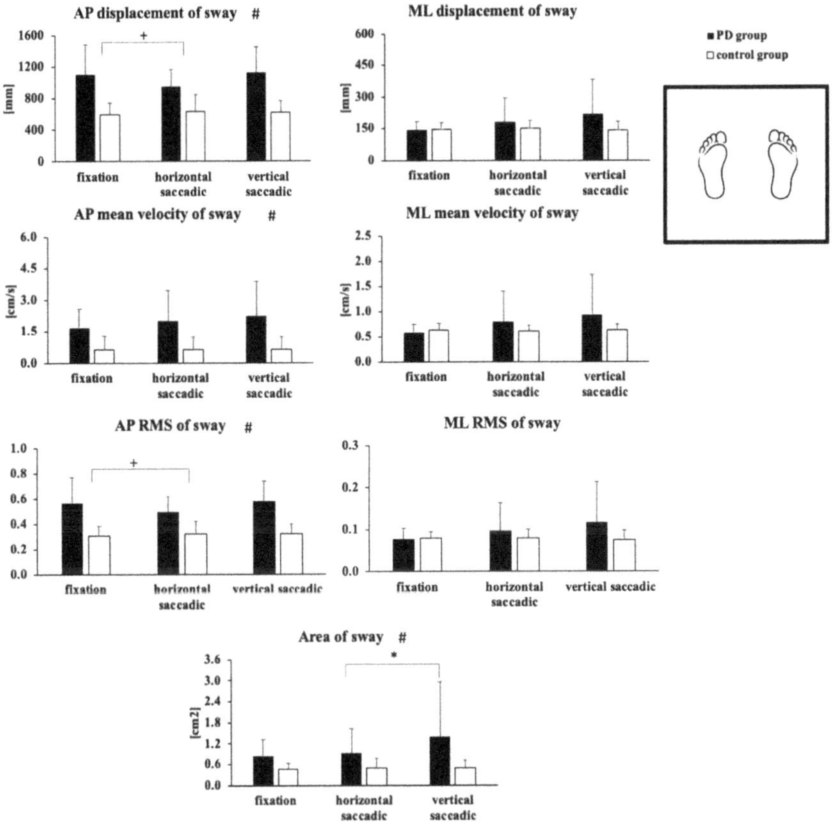

Figure 2. Means and standard deviations of CoP parameters for side-by-side stance during fixation and eye saccadic movements in horizontal and vertical directions in people with PD (PD group) and neurologically healthy individuals (control group). AP—anterior–posterior; ML—medial–lateral. # indicates significant group differences; + indicates significant differences between fixation condition and horizontal eye saccadic movements (both groups); * indicates significant differences between horizontal and vertical eye saccadic movements for the PD group.

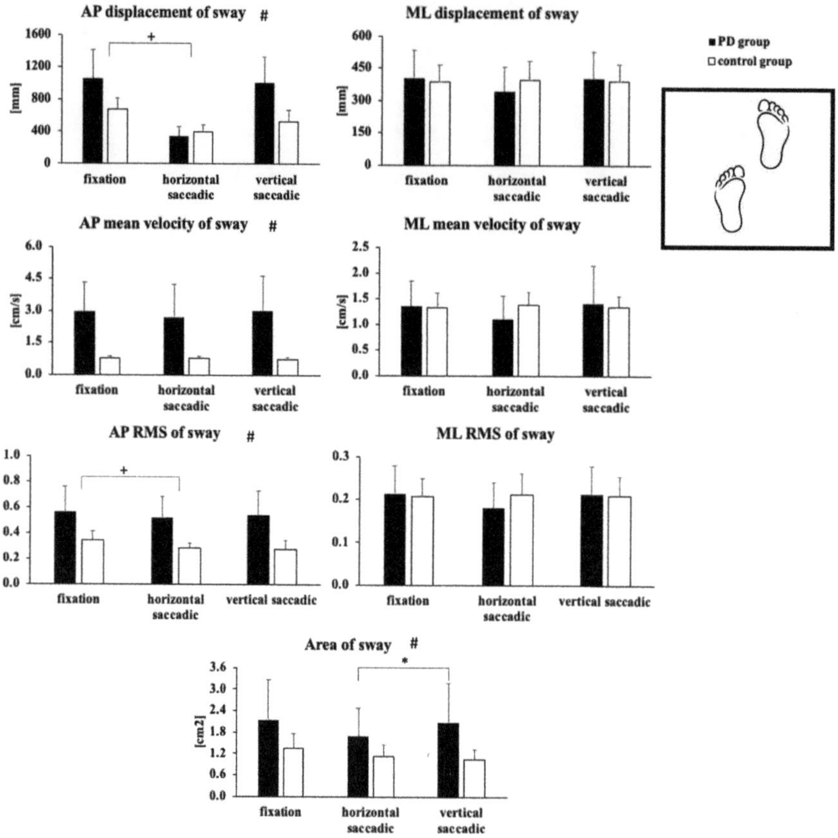

Figure 3. Means and standard deviations of CoP parameters for tandem stance during fixation and eye saccadic movements in horizontal and vertical directions in people with PD (PD group) and neurologically healthy individuals (control group). AP—anterior–posterior; ML—medial–lateral. # indicates significant group differences; + indicates significant differences between fixation condition and horizontal eye saccadic movements (both groups); * indicates significant differences between horizontal and vertical eye saccadic movements for the PD group.

ANOVAs indicated a significant group effect for AP displacement ($F_{1,18} = 24.23$, $p < 0.001$, $\eta^2 = 0.57$), mean velocity of sway ($F_{1,18} = 18.39$, $p < 0.001$, $\eta^2 = 0.50$) and RMS ($F_{1,18} = 24.96$, $p < 0.001$, $\eta^2 = 0.58$), and area of sway ($F_{1,18} = 9.80$, $p < 0.006$, $\eta^2 = 0.35$). The PD group showed higher levels of AP displacement, mean velocity of sway and RMS, and area of sway compared to the control group.

ANOVAs indicated a significant visual task effect for AP displacement ($F_{1,18} = 6.29$, $p < 0.005$, $\eta^2 = 0.25$) and RMS ($F_{1,18} = 6.76$, $p < 0.004$, $\eta^2 = 0.24$). During the horizontal saccadic movements, the participants in both groups reduced AP displacement and RMS compared to the fixation condition.

ANOVAs also indicated a significant base of support effect for ML displacement ($F_{1,18} = 64.49$, $p < 0.001$, $\eta^2 = 0.78$), mean velocity of sway ($F_{1,18} = 47.61$, $p < 0.001$, $\eta^2 = 0.72$) and RMS ($F_{1,18} = 61.21$, $p < 0.001$, $\eta^2 = 0.77$), AP mean velocity of sway ($F_{1,18} = 17.13$, $p < 0.01$, $\eta^2 = 0.48$), and area of sway ($F_{1,18} = 18.30$, $p < 0.001$, $\eta^2 = 0.50$). In both groups, the tandem stance resulted in higher levels of ML displacement and RMS, AP and ML mean velocity of sway, and area of sway than the side-by-side stance.

A group by base of support interaction was indicated in ANOVAs for AP mean velocity of sway ($F_{1,18} = 9.21$, $p < 0.007$, $\eta^2 = 0.33$). The PD group showed greater AP mean velocity of sway during the tandem stance compared to the side-by-side stance ($p < 0.001$), which was not found for the control group.

A group by visual task interaction was indicated in ANOVAs for the area of sway ($F_{2,36} = 3.59$, $p < 0.04$, $\eta^2 = 0.16$). Only for the PD group, the area of sway was higher when performing vertical saccadic eye movements compared to horizontal saccadic eye movements ($p < 0.005$), but there was no significant difference in comparison to fixation condition ($p = 0.124$).

3.2. Gaze Parameters

Figure 4 shows the gaze parameters for side-by-side and tandem stances during saccadic eye movements in horizontal and vertical directions. ANOVA did not reveal a significant interaction for (i) group and base of support, (ii) group and visual task, (iii) base of support and visual task, and (iv) group and base of support and visual task ($p > 0.05$).

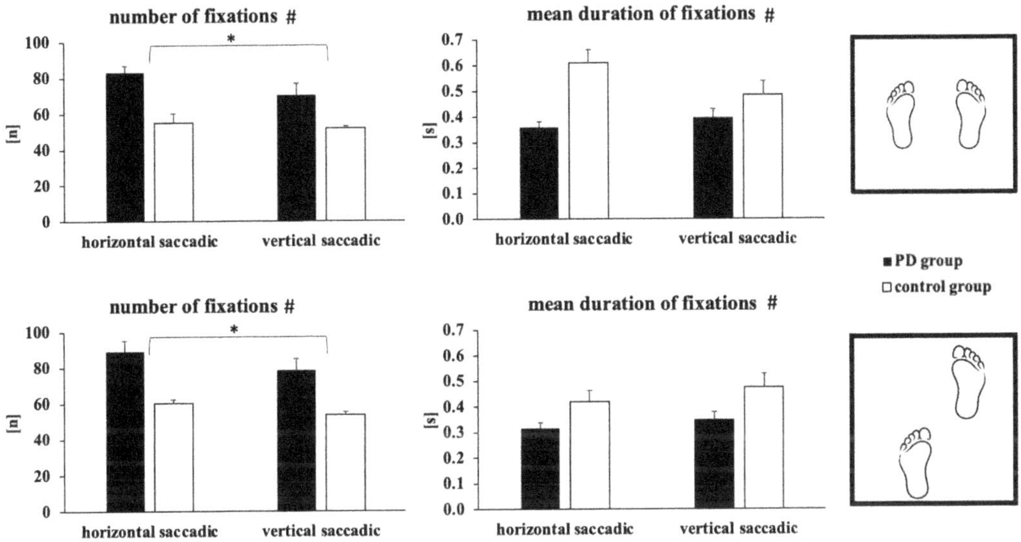

Figure 4. Means and standard errors of gaze parameters for side-by-side and tandem stances during fixation and eye saccadic movements in horizontal and vertical directions in people with PD (PD group) and neurologically healthy individuals (control group). # indicates significant group differences; * indicates significant differences between vertical and horizontal saccadic eye movements (both groups).

ANOVAs indicated a significant group effect for the number of fixations ($F_{1,17} = 24.60$, $p < 0.001$, $\eta^2 = 0.59$) and mean duration of fixations ($F_{1,18} = 5.19$, $p < 0.03$, $\eta^2 = 0.23$). In all tasks, the PD group exhibited a higher number of fixations and shorter mean duration of fixations compared to the control group.

ANOVAs indicated a significant visual task effect for the number of fixations ($F_{1,17} = 4.37$, $p < 0.05$, $\eta^2 = 0.21$). The participants in both groups exhibited a lower number of fixations in the vertical saccadic eye movements compared to horizontal saccadic conditions.

ANOVAs also indicated a significant base of support effect for the number of fixations ($F_{1,17} = 6.02$, $p < 0.02$, $\eta^2 = 0.26$). For both groups, the tandem stance resulted in a higher number of fixations than the side-by-side stance.

4. Discussion

We conducted a study to test whether people with PD are able to reduce their body sway when performing horizontal and vertical saccadic eye movements while standing with feet in a side-by-side or tandem stance. Our findings partially supported our first hypothesis, showing a reduction in body sway for both groups when performing horizontal saccadic eye movement compared to the eye fixation condition (we expected reduction only for neurologically healthy individuals). However, we did not find significant effects of the base of support on body sway during horizontal saccadic eye movements as we proposed in the first hypothesis. In addition, our second, third, and fourth hypotheses were fully supported by our findings: vertical saccadic eye movements did not reduce body sway for both people with PD and neurologically healthy individuals, gaze performance was poorer in people with PD compared to neurologically healthy individuals, especially during tandem stance, and vertical saccadic eye movements led to poorer gaze performance compared to horizontal saccadic eye movements, respectively. It is important to note that all significant effects had a large effect size, indicating that eye movements had a noteworthy impact on postural stability.

During horizontal saccadic eye movements, people with PD were able to reduce body sway during both side-by-side and tandem postural tasks by coupling eye movements and postural sway (Figures 2 and 3). This finding was unexpected and invalidated our main hypothesis. A posteriori, the literature also showed that neurologically affected populations are able to reduce their sway to perform gaze shift tasks as well as older healthy and younger populations [22–24]. The afferent and efferent mechanisms of visual stabilization of posture can explain these results [2]. On the one hand, the afferent mechanism involves minimizing changes in the projected image on the retina to maintain a relationship between visual information and body posture during fixation [23,29]. On the other hand, the efferent mechanism, particularly efference copy, attenuates body sway in an attempt to connect pre-saccadic and post-saccadic views of the scene, thus enhancing the spatial accuracy of the saccade concerning the target location [23,29]. Since the eye saccade condition requires greater postural stability to allow spatially more accurate gaze shifts, there seems to be a functional integration of postural and gaze control [1]. Also, similarly to neurologically healthy individuals, the achievement of a goal in a supra-postural task may explain the reduction in body sway during saccadic eye movements [27]. Therefore, there is an improvement in postural stability to facilitate, at least in part, the performance of this type of task [28].

One possible reason for the similar findings in people with PD, under dopaminergic medication, and healthy controls in our study (or other populations in other studies) could be related to subcortical control of posture. During horizontal saccadic eye movements, it is believed that the control of posture shifts to a more subcortical level [27]. It means that the brain uses the brainstem/cerebellum to control the body/head position, while other lower structures take control of eye movements. This hypothesis puts postural control on a "second goal", making it more "automatic" and reducing body sway during horizontal saccadic eye movement. Consistent with this argument, Bonnet et al. [37], Cruz et al. [7], and Feller et al. [38] showed that PD may not affect a person's ability to control their posture automatically in simple environments (e.g., when looking at a blank target). It means that people with PD are able to control their posture as well as healthy people in quiet stance. Furthermore, and based on our results, PD individuals are also able to improve their postural control, and thus show functional gaze and posture connection when performing horizontal saccadic eye movements. In other words, performing horizontal saccadic eye movements might be a useful strategy to help the central nervous system in people with PD to counter impairments in basal ganglia related to both gaze shift and postural control.

As expected, vertical saccadic eye movements did not reduce body sway and worsened gaze performance in both groups in comparison to the control fixation task. Therefore, people with PD did not show any functional coupling between gaze and posture but also did not show any impairment when performing vertical saccadic eye movements. As complementary information, vertical saccadic eye movements increased the area of sway in people with PD but only when contrasting the horizontal and vertical saccadic eye movements. Hence, this finding only shows that the coupling between gaze shift and postural control is significantly better when performing horizontal saccadic eye movements than vertical saccadic eye movements and not that there is a PD-related impairment in vertical saccadic eye movements (as again, there was no significant difference between body sway in vertical saccadic eye movements and fixation). We can provide some reasons to explain why the coupling between gaze shift and postural control was not functional when performing vertical saccadic eye movements.

Firstly, we need to acknowledge that the control of horizontal and vertical eye saccadic movements is distinct. In fact, on one hand, the cortical areas mainly involved in horizontal saccade generation are the contralateral parietal and frontal eye fields [26]. On the other hand, the circuits required to execute vertical saccades involve a higher level of activation in the right frontal eye field, cerebellar posterior lobe, and superior temporal gyrus [4]. Secondly, vertical eye movements are more variable compared to horizontal eye movements [25]. Our results seem to confirm that gaze performance is poorest during vertical eye movements. We found a lower number of fixations during vertical saccadic eye movements compared to horizontal saccadic conditions for both groups. However, we did not assess the variability of eye movements to fully confirm the existing literature. Thirdly, the direction of vertical gaze movement could have promoted slight up/down head rotations, which would have increased sway because the head is a heavy segment. Even if head rotations had been similar in left/right and up/down directions, they have clearly different effects on postural sway [39]. In fact, on one hand, in left/right head rotations, the head center of mass stays in line with the body center of mass, thus not creating any couple of rotations and ultimately not increasing postural sway. On the other hand, in up/down head rotations, the head center of mass is not aligned with the body center of mass anymore [39], thus creating couples of rotations that increase postural sway. Unfortunately, we cannot verify this a posteriori hypothesis, as we did not record head rotation. We can only suggest, from the results, that both groups were able to significantly reduce their sway when performing horizontal saccades but not vertical saccades. One relevant finding, though, is that due to the specificity of 90% in discriminating PD and healthy individuals, vertical saccades could be a biomarker for early diagnosis of PD [26] and postural impairments.

Our results validated the hypothesis that people with PD would have a worse gaze performance than neurologically healthy individuals. In fact, they revealed that people with PD used a higher number of fixations in all visual tasks in both postural tasks compared to healthy individuals, which could be an indication of reduced goal-directed control [40]. This suggests that people with PD have difficulty maintaining their visual attention and may shift their gaze toward irrelevant information, which could increase the number of fixations in the visual scene. This behavior can be dangerous, as shown by Gotardi et al. [41], who found that people with PD had an increased number of collisions while driving due to an overall increase in gaze fixations and a shift in visual attention toward task-irrelevant information. People with PD tend to have difficulties in processing temporal information, leading to deficits in temporal judgment [42]. According to Cruz et al. [43], individuals with PD exhibit delayed body sway to visual stimuli during continuous and predictable driving frequency compared to the control group, indicating some disruption in the visual–motor coupling. The pedunculopontine nucleus area, which is responsible for both saccades and posture functions, may explain some of the dysfunction seen in PD patients [20].

Although novel findings were presented above, some limitations of our study need to be discussed. First, the sample size was small but still fitted with the required minimum sample size. Additionally, all significant differences exhibited large effect sizes, which confirms that the number of participants in the study was adequate. A second limitation is that we refrained from the inclusion of patients in H&Y 4 and 5. So, our findings can be valid only for initial and moderate levels of PD. An extension of the study population to the late stages of PD would be relevant in future studies.

5. Conclusions

In conclusion, the novelty of our findings is that individuals with initial and moderate PD and ON-drug are able to couple eye movements and postural sway in order to improve body sway when saccadic horizontal eye movements are performed during standing tasks. Furthermore, these individuals with PD can reduce postural sway as efficiently as neurologically healthy people when performing horizontal saccades. However, vertical saccades could affect gaze behavior in PD, potentially compromising the effect of the efferent mechanism for postural stabilization in this population. Therefore, individuals with PD should better perform saccadic horizontal eye movements, and even exaggerate performing them, and they should be careful when performing saccadic vertical eye movements. Incorporating saccadic eye movements into postural training could be an effective intervention strategy for individuals with PD.

Author Contributions: Conceptualization, F.A.B., P.F.P. and S.T.R.; methodology, F.A.B., P.F.P., M.B.B. and S.T.R.; formal analysis, F.A.B., P.F.P., J.A.B., C.T.B., M.B.B. and S.T.R.; writing—original draft preparation, F.A.B., P.F.P., J.A.B., C.T.B., M.B.B. and S.T.R.; writing—review and editing, F.A.B., P.F.P., J.A.B., C.T.B., M.B.B. and S.T.R.; project administration, F.A.B., P.F.P. and S.T.R.; funding acquisition, F.A.B., P.F.P. and S.T.R. All authors have read and agreed to the published version of the manuscript.

Funding: This research was funded by São Paulo Research Foundation (FAPESP), grant numbers 2014/20549-0 and 2015/15928-4; Coordenação de Aperfeiçoamento de Pessoal de Nível Superior -Brasil (CAPES), grant number 001; National Council for Scientific and Technological Development (CNPq); CAPES-COFECUB, grant number Ma 1005/23 (ID Campus France: 49556YD) and 88881.712030/2022-01.

Institutional Review Board Statement: The study was conducted in accordance with the Declaration of Helsinki, and approved by the Ethics Committee of the School of Sciences at São Paulo State University (UNESP), Bauru (#11322/46/01/12 and 05/10/2012).

Informed Consent Statement: Informed consent was obtained from all subjects involved in the study.

Data Availability Statement: The data can be requested for the correspondent author.

Conflicts of Interest: The authors declare no conflicts of interest.

References

1. Stoffregen, T.A.; Bardy, B.G.; Bonnet, C.T.; Pagulayan, R.J. Postural Stabilization of Visually Guided Eye Movements. *Ecol. Psychol.* **2006**, *18*, 191–222. [CrossRef]
2. Guerraz, M.; Bronstein, A.M. Ocular versus extraocular control of posture and equilibrium. *Neurophysiol. Clin. Neurophysiol.* **2008**, *38*, 391–398. [CrossRef]
3. Ferrazzoli, D.; Fasano, A.; Maestri, R.; Bera, R.; Palamara, G.; Ghilardi, M.F.; Pezzoli, G.; Frazzitta, G. Balance Dysfunction in Parkinson's Disease: The Role of Posturography in Developing a Rehabilitation Program. *Parkinson's Dis.* **2015**, *2015*, 520128. [CrossRef] [PubMed]
4. Lemos, J.; Pereira, D.; Almendra, L.; Rebelo, D.; Patrício, M.; Castelhano, J.; Cunha, G.; Januário, C.; Cunha, L.; Freire, A.; et al. Distinct functional properties of the vertical and horizontal saccadic network in Health and Parkinson's disease: An eye-tracking and fMRI study. *Brain Res.* **2016**, *1648*, 469–484. [CrossRef] [PubMed]
5. Horak, F.B.; Dimitrova, D.; Nutt, J.G. Direction-specific postural instability in subjects with Parkinson's disease. *Exp. Neurol.* **2005**, *193*, 504–521. [CrossRef]
6. Rocchi, L.; Chiari, L.; Cappello, A.; Horak, F.B. Identification of distinct characteristics of postural sway in Parkinson's disease: A feature selection procedure based on principal component analysis. *Neurosci. Lett.* **2006**, *394*, 140–145. [CrossRef] [PubMed]
7. Cruz, C.F.; Piemonte, M.E.P.; Okai-Nobrega, L.A.; Okamoto, E.; de Souza Fortaleza, A.C.; Mancini, M.; Horak, F.B.; Barela, J.A. Parkinson's disease does not alter automatic visual-motor coupling in postural control. *Neurosci. Lett.* **2018**, *686*, 47–52. [CrossRef] [PubMed]
8. de Carvalho Costa, E.; Santinelli, F.B.; Moretto, G.F.; Figueiredo, C.; von Ah Morano, A.E.; Barela, J.A.; Barbieri, F.A. A multiple domain postural control assessment in people with Parkinson's disease: Traditional, non-linear, and rambling and trembling trajectories analysis. *Gait Posture* **2022**, *97*, 130–136. [CrossRef] [PubMed]

9. MMancini; Horak, F.B.; Zampieri, C.; Carlson-Kuhta, P.; Nutt, J.G.; Chiari, L. Trunk accelerometry reveals postural instability in untreated Parkinson's disease. *Park. Relat. Disord.* **2011**, *17*, 557–562. [CrossRef]
10. Doná, F.; Aquino, C.C.; Gazzola, J.M.; Borges, V.; Silva, S.M.C.A.; Ganança, F.F.; Caovilla, H.H.; Ferraz, H.B. Changes in postural control in patients with Parkinson's disease: A posturographic study. *Physiotherapy* **2016**, *102*, 272–279. [CrossRef]
11. Palakurthi, B.; Burugupally, S.P. Postural Instability in Parkinson's Disease: A Review. *Brain Sci.* **2019**, *9*, 239. [CrossRef]
12. Allen, N.E.; Schwarzel, A.K.; Canning, C.G. Recurrent Falls in Parkinson's Disease: A Systematic Review. *Parkinson's Dis.* **2013**, *2013*, 906274. [CrossRef] [PubMed]
13. Beretta, V.S.; Barbieri, F.A.; Orcioli-Silva, D.; Santos, P.C.R.D.; Simieli, L.; Vitório, R.; Gobbi, L.T.B. Can Postural Control Asymmetry Predict Falls in People with Parkinson's Disease? *Mot. Control* **2018**, *22*, 449–461. [CrossRef] [PubMed]
14. Beretta, V.S.; Gobbi, L.T.B.; Lirani-Silva, E.; Simieli, L.; Orcioli-Silva, D.; Barbieri, F.A. Challenging postural tasks increase asymmetry in patients with Parkinson's disease. *PLoS ONE* **2015**, *10*, e0137722. [CrossRef]
15. Mancini, M.; Carlson-Kuhta, P.; Zampieri, C.; Nutt, J.G.; Chiari, L.; Horak, F.B. Postural sway as a marker of progression in Parkinson's disease: A pilot longitudinal study. *Gait Posture* **2012**, *36*, 471–476. [CrossRef]
16. Takakusaki, K. Functional Neuroanatomy for Posture and Gait Control. *J. Mov. Disord.* **2017**, *10*, 1–17. [CrossRef] [PubMed]
17. Brown, L.A.; Cooper, S.A.; Doan, J.B.; Dickin, D.C.; Whishaw, I.Q.; Pellis, S.M.; Suchowersky, O. Parkinsonian deficits in sensory integration for postural control: Temporal response to changes in visual input. *Park. Relat. Disord.* **2006**, *12*, 376–381. [CrossRef]
18. Terao, Y.; Fukuda, H.; Ugawa, Y.; Hikosaka, O. New perspectives on the pathophysiology of Parkinson's disease as assessed by saccade performance: A clinical review. *Clin. Neurophysiol.* **2013**, *124*, 1491–1506. [CrossRef]
19. Anderson, T.J.; MacAskill, M.R. Eye movements in patients with neurodegenerative disorders. *Nat. Rev. Neurol.* **2013**, *9*, 74–85. [CrossRef]
20. Ewenczyk, C.; Mesmoudi, S.; Gallea, C.; Welter, M.-L.; Gaymard, B.; Demain, A.; Cherif, L.Y.; Degos, B.; Benali, H.; Pouget, P.; et al. Antisaccades in Parkinson disease. *Neurology* **2017**, *88*, 853–861. [CrossRef]
21. Beh, S.C.; Frohman, T.C.; Frohman, E.M. Cerebellar Control of Eye Movements. *J. Neuro-Ophthalmol.* **2017**, *37*, 87–98. [CrossRef] [PubMed]
22. Rodrigues, S.T.; Polastri, P.F.; Carvalho, J.C.; Barela, J.A.; Moraes, R.; Barbieri, F.A. Saccadic and smooth pursuit eye movements attenuate postural sway similarly. *Neurosci. Lett.* **2015**, *584*, 292–295. [CrossRef] [PubMed]
23. Rodrigues, S.T.; Aguiar, S.A.; Polastri, P.F.; Godoi, D.; Moraes, R.; Barela, J.A. Effects of saccadic eye movements on postural control stabilization. *Mot. Rev. Educ. Física* **2013**, *19*, 614–619. [CrossRef]
24. Aguiar, S.A.; Polastri, P.F.; Godoi, D.; Moraes, R.; Barela, J.A.; Rodrigues, S.T. Effects of saccadic eye movements on postural control in older adults. *Psychol. Neurosci.* **2015**, *8*, 19–27. [CrossRef]
25. Polastri, P.F.; Barbieri, F.A.; Brito, M.B.; Bonfim, J.V.A.A.; Gotardi, G.C.; Kuga, G.K.; Rodrigues, S.T. Semi tandem base of support degrades both saccadic gaze control and postural stability particularly in older adults. *Neurosci. Lett.* **2019**, *705*, 227–234. [CrossRef] [PubMed]
26. Waldthaler, J.; Tsitsi, P.; Svenningsson, P. Vertical saccades and antisaccades: Complementary markers for motor and cognitive impairment in Parkinson's disease. *NPJ Park. Dis.* **2019**, *5*, 11. [CrossRef] [PubMed]
27. Stoffregen, T.A.; Smart, L.J.; Bardy, B.G.; Pagulayan, R.J. Postural stabilization of looking. *J. Exp. Psychol. Hum. Percept. Perform.* **1999**, *25*, 1641–1658. [CrossRef]
28. Oullier, O.; Bardy, B.G.; Stoffregen, T.A.; Bootsma, R.J. Postural coordination in looking and tracking tasks. *Hum. Mov. Sci.* **2002**, *21*, 147–167. [CrossRef] [PubMed]
29. Kowler, E. Eye movements: The past 25 years. *Vis. Res.* **2011**, *51*, 1457–1483. [CrossRef]
30. Hughes, A.J.; Daniel, S.E.; Kilford, L.; Lees, A.J. Accuracy of clinical diagnosis of idiopathic Parkinson's disease: A clinico-pathological study of 100 cases. *J. Neurol. Neurosurg. Psychiatry* **1992**, *55*, 181–184. [CrossRef]
31. Goetz, C.G.; Poewe, W.; Rascol, O.; Sampaio, C.; Stebbins, G.T.; Counsell, C.; Giladi, N.; Holloway, R.G.; Moore, C.G.; Wenning, G.K.; et al. Movement Disorder Society Task Force report on the Hoehn and Yahr staging scale: Status and recommendations. *Mov. Disord.* **2004**, *19*, 1020–1028. [CrossRef] [PubMed]
32. Almeida, O.P. The Mini-Mental State Examination and the Diagnosis of Dementia in Brazil. *Arq. Neuropsiquiatr.* **1998**, *56*, 605–612. [CrossRef]
33. Goetz, C.G.; Fahn, S.; Martinez-Martin, P.; Poewe, W.; Sampaio, C.; Stebbins, G.T.; Stern, M.B.; Tilley, B.C.; Dodel, R.; Dubois, B.; et al. Movement Disorder Society-sponsored revision of the Unified Parkinson's Disease Rating Scale (MDS-UPDRS): Process, format, and clinimetric testing plan. *Mov. Disord.* **2007**, *22*, 41–47. [CrossRef]
34. Araújo-Silva, F.; Santinelli, F.B.; Imaizumi, L.F.I.; Silveira, A.P.B.; Vieira, L.H.P.; Alcock, L.; Barbieri, F.A. Temporal dynamics of cortical activity and postural control in response to the first levodopa dose of the day in people with Parkinson's disease. *Brain Res.* **2022**, *1775*, 147727. [CrossRef] [PubMed]
35. Barbieri, F.A.; Simieli, L.; Orcioli-Silva, D.; Baptista, A.M.; Pestana, M.B.; Beretta, V.S.; Santos, P.C.R.D.; Gobbi, L.T.B. Obstacle Avoidance Increases Asymmetry of Crossing Step in Individuals with Parkinson's Disease and Neurologically Healthy Individuals. *J. Mot. Behav.* **2018**, *50*, 17–25. [CrossRef]
36. Richardson, J.T.E. Eta squared and partial eta squared as measures of effect size in educational research. *Educ. Res. Rev.* **2011**, *6*, 135–147. [CrossRef]

37. Bonnet, C.T.; Delval, A.; Singh, T.; Kechabia, Y.-R.; Defebvre, L. New insight into Parkinson's disease-related impairment of the automatic control of upright stance. *Eur. J. Neurosci.* **2020**, *52*, 4851–4862. [CrossRef]
38. Feller, K.J.; Peterka, R.J.; Horak, F.B. Sensory Re-weighting for Postural Control in Parkinson's Disease. *Front. Hum. Neurosci.* **2019**, *13*, 126. [CrossRef] [PubMed]
39. Fukushima, J.; Asaka, T.; Fukushima, K. Postural changes during eye–head movements. *Prog. Brain Res.* **2008**, *171*, 335–338. [CrossRef]
40. Derakshan, N.; Ansari, T.L.; Hansard, M.; Shoker, L.; Eysenck, M.W. Anxiety, Inhibition, Efficiency, and Effectiveness. *Exp. Psychol.* **2009**, *56*, 48–55. [CrossRef]
41. Gotardi, G.C.; Barbieri, F.A.; Simão, R.O.; Pereira, V.A.; Baptista, A.M.; Imaizumi, L.F.; Moretto, G.; Navarro, M.; Polastri, P.F.; Rodrigues, S.T. Parkinson's disease affects gaze behaviour and performance of drivers. *Ergonomics* **2022**, *65*, 1302–1311. [CrossRef]
42. Lucas, M.; Chaves, F.; Teixeira, S.; Carvalho, D.; Peressutti, C.; Bittencourt, J.; Velasques, B.; Menéndez-González, M.; Cagy, M.; Piedade, R.; et al. Time perception impairs sensory-motor integration in Parkinson's disease. *Int. Arch. Med.* **2013**, *6*, 39. [CrossRef]
43. Cruz, C.F.; Barela, A.M.F.; Doná, F.; Genoves, G.G.; Ferraz, H.B.; de Azevedo Silva, S.M.C.; Barela, J.A. Parkinson's disease delays predictable visual cue processing without affecting complex and unpredictable visual cue processing in postural control. *Brain Res.* **2021**, *1751*, 147209. [CrossRef]

biomechanics

MDPI

Article

Evolution of Gait Biomechanics During a Nine-Month Exercise Program for Parkinson's Disease: An Interventional Cohort Study

Dielise Debona Iucksch [1], Elisangela Ferretti Manffra [2,*] and Vera Lucia Israel [3]

1 Postgraduate Program in Physical Education, Federal University of Paraná, Curitiba 81531-980, Brazil; dielise@ufpr.br
2 Health Technology Graduate Program, Pontifícia Universidade Católica do Paraná, Curitiba 80215-901, Brazil
3 Department of Physical Therapy Prevention and Rehabilitation, Federal University of Paraná, Curitiba 81531-980, Brazil; vera.israel@ufpr.br
* Correspondence: elisangela.manffra@pucpr.br

Abstract: It is well established that combining exercise with medication may benefit functionality in individuals with PD (Parkinson's disease). However, the long-term evolution of gait biomechanics under this combination remains poorly understood. Objectives: This study aims to analyze the evolution of spatiotemporal gait parameters, kinetics, and kinematics throughout a long-term exercise program conducted in water and on dry land. Methods: We have compared the trajectories of biomechanical variables across the treatment phases using statistical parametric mapping (SPM). A cohort of fourteen individuals with PD (mean age: 65.6 ± 12.1 years) participated in 24 sessions of aquatic exercises over three months, followed by a three-month retention phase, and then 24 additional sessions of land-based exercises. Three-dimensional gait data and spatiotemporal parameters were collected before and after each phase. Two-way ANOVA with repeated measures was used to compare spatiotemporal parameters. Results: The walking speed increased while the duration of the double support phase decreased. Additionally, the knee extensor moment consistently increased in the entire interval from midstance to midswing (20% to 70% of the stride period), approaching normal gait patterns. Regarding kinematics, significant increases were observed in both hip and knee flexion angles. Furthermore, the abnormal ankle dorsiflexion observed at the foot strike disappeared. Conclusions: These findings collectively suggest positive adaptations in gait biomechanics during the observation period.

Keywords: Parkinson's disease; gait biomechanics; kinematics; kinetics; exercise program; statistical parametric mapping

1. Introduction

Parkinson's disease (PD) is a multifaceted and intricate neurodegenerative condition [1] characterized by a spectrum of motor and non-motor symptoms that significantly impact independence and quality of life [2]. Among these symptoms, gait abnormalities stand out as one of the most prevalent and disabling motor impairments [3].

In PD, specific gait dysfunctions manifest as reductions in speed, step and stride lengths, swing time, and range of motion [3,4]. Additionally, alterations in the timing and amplitude of leg muscle activation are observed [5,6]. Ankle angle kinematics are often notably affected, alongside reductions in joint moments and power generation during pre-swing [7–10]. These gait impairments significantly contribute to the functional limitations experienced by individuals with PD.

The primary treatment for motor dysfunction in individuals with PD is medication. However, compelling evidence suggests that combining it with physical exercise may significantly enhance quality of life and motor outcomes, particularly those related to gait [11–13]. Consequently, physical therapy guidelines have recommended physical exercises for PD patients [12,14,15].

Several studies on gait function have indicated improvements in functional and spatiotemporal gait outcomes following physical exercise interventions [12]. For example, Rafferty et al. [16] compared the effects of progressive resistance exercise and a multimodal exercise program on spatiotemporal and stability-related gait outcomes, noting enhanced off-medication gait velocity and cadence in both groups. Shen et al. [17] demonstrated that balance and gait training with augmented feedback enhanced gait velocity and stride length, whereas the active control group showed improvement in gait velocity only. Additionally, individuals with PD who participated in aquatic gait training significantly improved spatiotemporal parameters [6].

In a specific guideline for exercise protocols targeting gait function, Ni et al. [12] recommended a comprehensive approach that includes multidimensional physical therapy (balance and gait training), treadmill gait training with body weight support, cycling, aquatic exercises, resistance training, and complementary treatments such as tai chi, yoga, and tango. The authors endorsed various modalities because no single approach was deemed superior to the others when considering overall gait function outcomes.

More recently, systematic reviews with meta-analyses have highlighted the superiority of aquatic exercises (AEs) for enhancing functional mobility and balance [11,13,18,19]. Hvingelby et al. [13] concluded that aquatic therapy with dual-task training exhibited the most significant effect on dynamic gait outcomes, such as scores in Timed Up and Go, Dynamic Gait, and other gait functional scales, compared to other exercise protocols. Conversely, Ernst et al. [11] concluded that aqua-based exercises were superior to other forms of exercise for improving quality of life and were equivalent to them regarding motor signs.

In summary, it is well established that both water-based and land-based physical exercises can improve functional gait outcomes in PD. However, studies have primarily focused on spatiotemporal parameters or functional mobility scales [20], overlooking the biomechanics of gait. One study has described changes in gait kinematics and muscle recruitment patterns following aquatic training protocols [6,21], but, to our knowledge, none have explored gait kinetics. Therefore, a gap remains in understanding how gait kinetics evolve in individuals with PD undergoing interventions, especially aquatic ones, over the long term.

While studies analyzing gait kinetics in PD after AE training are lacking, it has been demonstrated that healthy individuals exhibit different joint moments when walking underwater compared to on land [22,23]. On the other hand, a similar set of muscle synergies has been observed in both land and water walking [24]. Thus, training gait underwater may induce force production and muscle activation patterns that could influence gait on land.

More recently, a study analyzed the gait kinetics of individuals with PD in comparison with healthy individuals during walking and obstacle crossing, both with and without the use of virtual reality [25]. They observed a difference between groups regarding the maximum joint moments, especially in the sagittal plane, and also that virtual reality influenced the kinetics of both groups. Therefore, joint moments in PD might bring valuable information.

Given this scenario, this study aims to analyze the evolution of gait biomechanics throughout a long-term exercise program conducted in water and on dry land. We hypothesize that angular kinematics and kinetics will exhibit changes across time, related to a more functional gait.

Usually, studies on gait biomechanics in PD investigate the behavior of discrete parameters such as maximum, minimum, or average values during the step, stride, or some gait phase [26]. To provide a more comprehensive view of the phenomenon, we decided not to compare discretized values, but rather the entire trajectories of kinematic and kinetic variables along the stride. To do so, we employed the statistical parametric mapping (SPM) analysis [27], which also reduces the risk of bias introduced by the choice of specific discrete parameters.

2. Materials and Methods

This is a longitudinal interventional cohort study conducted over nine months during which individuals with Parkinson's disease participated in a physical exercise program.

This study spanned over nine months and included assessment and intervention sessions, as depicted in Figure 1.

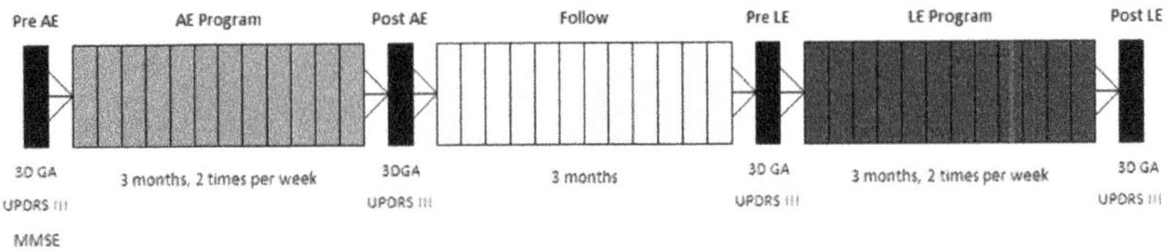

Figure 1. Timeline of exercise programs and assessments. Abbreviations: Pre AE, Post AE, Pre LE, and Post LE stand for assessment sessions, pre-aquatic exercise, post-aquatic exercise, pre-land exercise, and post-land exercise, respectively; 3D GA, three-dimensional gait analysis; UPDRS, Unified Parkinson's Disease Rating Scale; and MMSE, Mini-Mental State Examination.

To ensure assessments and interventions were conducted during the "on" phase of medication, all sessions took place 1 to 2 h after taking dopaminergic medication. The assessment sessions were scheduled to match each participant's usual medication routine. Since the exercise sessions were conducted in groups, some participants had to make minor adjustments to their medication schedules, with the counseling of their physicians.

2.1. Participants

Nineteen individuals diagnosed with Parkinson's disease were initially recruited from the local association, the Parkinson Paraná Association. Four subjects withdrew from this study during the intervention phase, and one subject was excluded from the statistical analysis due to corrupted biomechanical data. We have conducted a complete-case analysis, i.e., only data from 14 participants with Parkinson's disease (9 men) were analyzed.

The participants' average age, height, and weight were 65.6 ± 12.1 years, 165.4 ± 9.6 cm, and 70.5 ± 10.9 kg, respectively. The average score on the modified Hoehn and Yahr (H&Y) scale was 2.8 ± 1.1, with 8 participants classified in stages I–II and 6 in stages III–IV. The average score on the Unified Parkinson's Disease Rating Scale (UPDRS) part III was 14.4 ± 6.1 points, and on the Mini-Mental State Examination, it was 27.1.

This study's inclusion criteria were a confirmed diagnosis of idiopathic Parkinson's disease, an age between 40 and 90, the ability to walk unassisted, a Mini-Mental State Examination score equal to or above 20, and medical clearance to engage in physical exercise both on land and in a heated pool. Additionally, participants were required to maintain their regular physical activity unchanged throughout the study duration.

Exclusion criteria included the presence of other neurological conditions that affect gait or cognitive function, adjustments in the dosage of Parkinson's disease medications during the study period, and failure to attend assessment sessions or participate in more than 10% of the exercise program sessions.

2.2. Assessment Sessions—Data Collection

Assessment sessions included the administration of clinical scales and the collection of biomechanical data.

Clinical scales included the Mini-Mental State Examination, part III of the Unified Parkinson's Disease Rating Scale (UPDRS) [28], and the modified Hoehn and Yahr (H&Y) scale [29].

Biomechanical data were collected using a 3D motion tracking system equipped with eleven infrared cameras (Vicon Motion System Ltd., Oxford, UK) and a force plate (Advanced Mechanical Technology Inc., Watertown, USA). Fifteen reflective markers were positioned on anatomical landmarks in accordance with the plug-in gait lower body model [30].

Participants were instructed to walk along a 9 m pathway at a self-selected speed. Each session consisted of one trial of walking back and forth along the pathway for familiarization, followed by at least ten recording trials. Participants could take breaks during the session if needed to prevent fatigue.

The markers were sampled at 100 Hz, while the ground reaction forces (GRFs) were sampled at 1000 Hz. The Vicon Nexus® 2.5 software was used for raw data acquisition, as well as for the calculation of joint angles and moments. At least three gait cycles were analyzed for each participant and limb. After conducting consistency analyses, data from one stride of the most compromised lower limb and one stride of the least compromised lower limb for each subject were selected for further analysis of gait variables. We defined the most compromised limb as the one that first exhibited symptoms of the disease.

Customized signal processing scripts were developed in MATLAB R2018a (The Mathworks Inc., Natick, MA, USA) for filtering using a fourth-order zero-lag Butterworth low-pass filter with a 6 Hz cut-off frequency and time normalization to 100% gait cycle. Joint moments (measured in Nm) were normalized to the participant's body mass (measured in kilograms).

2.3. Exercise Programs

Before starting the 12-week aquatic exercise (AE) program, the volunteers participated in two sessions to familiarize themselves with the environment. The water temperature was kept at approximately 33 °C. Following a 12-week retention period, participants began the 12-week land environment (LE) program. The sessions took place twice a week in both environments and lasted about 50 min.

The choice of 12-week interventions was based on the findings of Carroll et al. [31], and the idea of including both water and land exercises stemmed from clinical practice, where exercises are seldom performed solely in the AE. They typically co-occur in both environments or sequentially in alternating environments, as in this study.

The duration of the retention period was chosen to match that of the exercise programs. During that time, participants were instructed not to enroll in any other exercise programs or physical activities, except those they were already enrolled in during the intervention period. They were also instructed to maintain their habitual medication regimen. This was monitored through self-report and confirmed during follow-up contacts.

The exercise program was designed in accordance with GRADE recommendations for the core domains of gait and balance, as outlined in the European PD Guidelines [14]. Its structure and content were aligned with multidimensional physical therapy recommendations for improving gait and balance in individuals with Parkinson's disease [12].

It encompassed multimodal exercises to maximize intervention effects. Each session included a warm-up, strength and power training exercises for the lower limbs, balance training, and a cool-down (Ai Chi or Tai Chi Chuan). We provide a brief overview of the program in the following paragraphs, and more details can be found in [32,33].

The warm-up in AE and LE included walking combined with coordinated movements of the upper limbs, trunk, and head. These were performed in various directions and at different speeds, incorporating turns, squats, skipping, and running. It involved walking forward, backward, and sideways at a comfortable pace, as well as performing hip and knee flexion with alternating lower limbs while rotating the trunk, using a kickboard (in AE) or a stick (in AE). Participants also ran or walked as fast as possible. Different instructions were provided in each environment to ensure proper movement execution. The warm-up lasted about 10 min.

The strength and power training was based on the protocol of Kanitz et al. [34] and consisted of kicks executed in the directions of hip and knee flexion/extension, as well as hip adduction/abduction. It progressed over the weeks by gradually increasing load, movement speed, number of repetitions, quality of execution, and overall difficulty [14]. During the first four weeks, the exercises in AE and LE consisted of two repetitions of 20 s of knee flexion–extension and hip abduction–adduction with each lower limb, without external resistance. From the 5th to the 8th week, the protocol progressed to three repetitions with external resistance provided by flotation devices and elastic bands in AE and LE, respectively. In the last four weeks, the protocol advanced to four 15 s repetitions with a 1 min and 30 s rest interval between them. Participants were always instructed to perform as many kicks as possible within the allotted time, aiming for the largest possible range of motion. The intensity of each repetition was monitored using the Borg Rating of Perceived Exertion Scale—CR20. The participants were instructed to report their perceived effort during the rest intervals between repetitions. For this study, the target intensity range was set between 13 and 17 points, corresponding approximately to 66% and 80% of the maximum voluntary force output, respectively, as recommended by the American College of Sports Medicine [35,36]. Verbal incentives were provided accordingly to maintain the target intensity.

Balance exercises consisted of walking with reduced base of support, maintaining challenging postures (e.g., far-reaching in unipodal support, standing on an unstable surface such as trampolines), and other motor tasks requiring postural control. New balance tasks were introduced every four weeks to increase complexity progressively, reduce reliance on upper limb support, and further narrow the base of support.

The cool-down consisted of Ai Chi or Tai Chi Chuan exercises in AE or LE, respectively. The Ai Chi exercises involve concentrating on breathing and performing slow movements of elevation of MMII, floating, and simply remaining calm, contemplating [37]. The Tai Chi Chuan movements were Wuchi/Tai Chi Opening, Carelessly Rolling Up the Sleeves, Open and Close the Hands, Single Whip, and Cloud Hands [38].

To ensure standardization of the exercise protocol across participants, the sessions were delivered to groups of participants by the same team of trained instructors throughout the programs.

2.4. Statistical Analyses

To compare spatiotemporal parameters across assessments, we employed a two-way repeated measures ANOVA. Normality distribution was assessed using the Shapiro–Wilk test and sphericity with the Mauchly test. In cases where sphericity was violated, a

Greenhouse–Geisser correction was applied. Post hoc pairwise comparisons between assessments were conducted using the Bonferroni test. These statistical analyses were conducted using IBM SPSS Statistics (version 20), with statistical significance set at $p < 0.05$. Effect sizes were determined using partial eta squared (ηp^2) and Cohen's d, with ηp^2 values classified as small (<0.06), moderate (>0.06 to 0.14), and large (>0.14) following Cohen's guidelines [32] and Cohen's, as small (<0.2), moderate (>0.2 to 0.5), medium (>0.5 to 0.8), and large (>0.8) [39]. The power (1-β) was estimated using the software G*Power v3.1.9.7, with the following settings: test family, ANOVA repeated measures within factors; type of power analysis, post hoc; effect size as in SPSS v.20; total sample size, 14; number of groups, 1; number of measurements, 4; and non-sphericity correction, 1.

To compare joint kinetics and kinematics trajectories across assessments and between limbs, we employed one-dimensional statistical parametric mapping (SPM) analysis [27,40].

We chose SPM because it enables us to compare the whole trajectories, thereby avoiding possible bias introduced by the choice of one or another discretized measure. Moreover, SMP allows for identifying the periods in time where trajectories might differ, and not only if they differ or not.

The method consists of generating a statistical parametric map by computing univariate F- or t-statistics at each point of the trajectories, namely SPM{F}, SPM{t}, or SnPM{t} for the ANOVA, parametric t-test, or non-parametric t-test, respectively [27]. Subsequently, random field theory is used to estimate the critical threshold above which only a percentage (e.g., 5% for $\alpha = 0.05$) of random data is expected to exceed.

We conducted SPM analysis in MATLAB R2018a (The Mathworks Inc., Natick, MA, USA) using the open-source SPM1d code version M.0.4.7 (https://spm1d.org/) (accessed on 15 July 2025). We used the function "spm1d.stats.anova2onerm" to conduct a two-way ANOVA with two factors—limb and assessment—with repeated measures on factor assessment, and the function "spm.inference" with the significance level set to 0.05 to define the threshold in SPM{F}.

When ANOVA indicated significant differences, we used parametric or non-parametric paired t-tests to compare the assessments pairwise using the functions "spm1d.stats.ttest_paired" or "spm1d.stats.nonparam.ttest_paired", respectively, and the function "spm.inference" with significance threshold set to 0.008 to account for the Bonferroni correction. This significance level is the ratio of $\alpha = 0.05$ divided by six, corresponding to the number of pairwise comparisons.

3. Results

The spatiotemporal parameters are presented in Table 1.

Table 1. Comparison of the spatiotemporal variables across assessments. Values are expressed as mean ± standard deviation; the *p*-values from the post hoc test are indicated along with Cohen's d effect size when there was statistical significance.

Condition	Aquatic Exercise Program			Land Exercise Program			Whole Program (AE, Follow Up, LE)		
	Pre AE	Post AE	*p*-Value Cohen's d	Pre LE	Post LE	*p*-Value Cohen's d	Pre AE	Post LE	*p*-Value Cohen's d
Gait Speed (m/s)	1.0 ± 0.2	1.1 ± 0.1	$p = 0.30$	1.1 ± 0.2	1.2 ± 0.2	$p = 0.11$	1.0 ± 0.2	1.2 ± 0.2	$p = 0.02$ * d = 0.71
Cadence (steps/min)	108.7 ± 10.3	108.1 ± 6.3	$p = 0.78$	110.4 ± 8.5	114.9±10.0	$p < 0.01$ * d = 0.48	108.7 ± 10.3	114.9 ± 10.0	$p < 0.01$ * d = 0.61
Stride Length (cm)	1.1 ± 0.2	1.2 ± 0.1	$p = 0.97$	1.2 ± 0.1	1.2 ± 0.2	$p = 0.87$	1.13±0.16	1.21±0.16	$p = 0.07$
Swing Time (%)	38.0 ± 2.7	38.0 ± 2.4	$p = 0.97$	37.7 ± 2.5	38.7 ± 2.9	$p = 0.03$ * d = 0.37	38.0 ± 2.7	38.7 ± 2.9	$p = 0.21$
Double Support (%)	25.0 ± 3.5	23.9 ± 3.1	$p = 0.20$	24.0 ± 3.5	23.0 ± 3.6	$p = 0.03$ * d = 0.29	25.0 ± 3.5	23.0 ± 3.6	$p < 0.01$ * d = 0.56

Abbreviations: Pre AE, pre-aquatic exercise assessment; Post AE, post-aquatic exercise assessment; Pre LE, pre-land exercise assessment; Post LE, post-land exercise assessment. (*) indicates statistical significance.

After the entire observation period, walking speed increased significantly [F (1, 13) = 6.11, p = 0.02, ηp^2 = 0.32, (1-β) = 0.33]. Post-LE assessments revealed significant increases in cadence [F (1, 13) = 22.97, p < 0.01, ηp^2 = 0.63, (1-β) = 0.71], which were sustained throughout subsequent assessments [F (1, 13) = 9.58, p < 0.01, ηp^2 = 0.42, (1-β) = 0.28]. Swing time also significantly improved in the post-LE phase [F (1, 13) = 5.33, p = 0.03, ηp^2 = 0.29, (1-β) = 0.20]. Double support time showed improvement post-LE [F (1, 13) = 5.65, p = 0.03, ηp2 = 0.30, (1-β) = 21], and throughout the entire program duration [F (1, 13) = 10.2, p < 0.01, ηp^2 = 0.44, (1-β) = 0.32]. However, there were no significant differences between the post-LE and pre-AE assessments for the other spatiotemporal variables.

The two-way ANOVA with statistical parametric mapping (SPM) analysis indicated no significant difference between limbs regarding angle kinetics (Figure 2, Main Factor A) or kinematics (Figure 3, Main Factor A). However, differences were observed across assessments for all joints concerning kinetics (Figure 2, Main Factor B) and kinematics (Figure 3, Main Factor B).

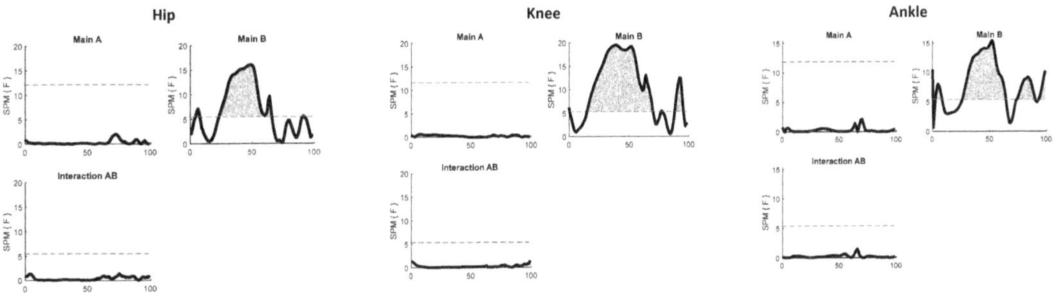

Figure 2. Results of the SPM two-way repeated measures ANOVA for hip, knee, and ankle joint moments. Main A: main factor A corresponds to the most and least compromised lower limbs; Main B: main factor B corresponds to assessments (Pre AE, Post AE, Pre LE, and Post LE). Grey areas above or below the red dotted lines indicate a significant difference (p < 0.05). Abbreviations: Pre AE, pre-aquatic exercise assessment; Post AE, post-aquatic exercise assessment; Pre LE, pre-land exercise assessment; Post LE, post-land exercise assessment.

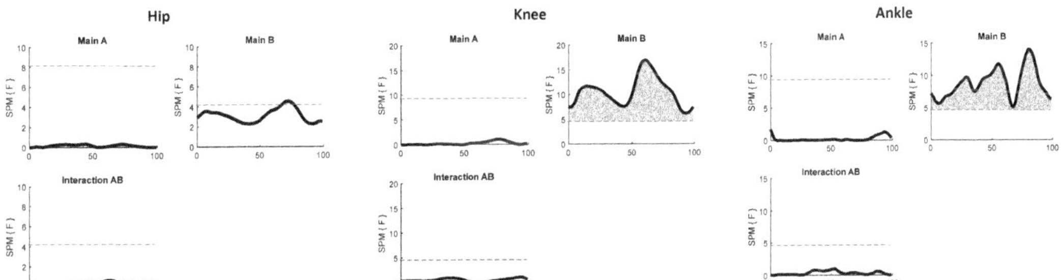

Figure 3. Result of the SPM two-way repeated measures ANOVA analysis for hip, knee, and ankle joint angles. Main A: main factor A corresponds to the most and least compromised lower limbs; Main B: main factor B corresponds to assessments (Pre AE, Post AE, Pre LE, and Post LE). Grey areas above or below the red dotted lines indicate a significant difference (p < 0.05). Abbreviations: Pre AE, pre-aquatic exercise assessment; Post AE, post-aquatic exercise assessment; Pre LE, pre-land exercise assessment; Post LE, post-land exercise assessment.

The SPM analysis revealed a difference only between pre-AE and post-LE. Figures 4 and 5 illustrate the trajectories of joint kinetics and kinematics before AE and after LE, along with the t-values calculated via SPM.

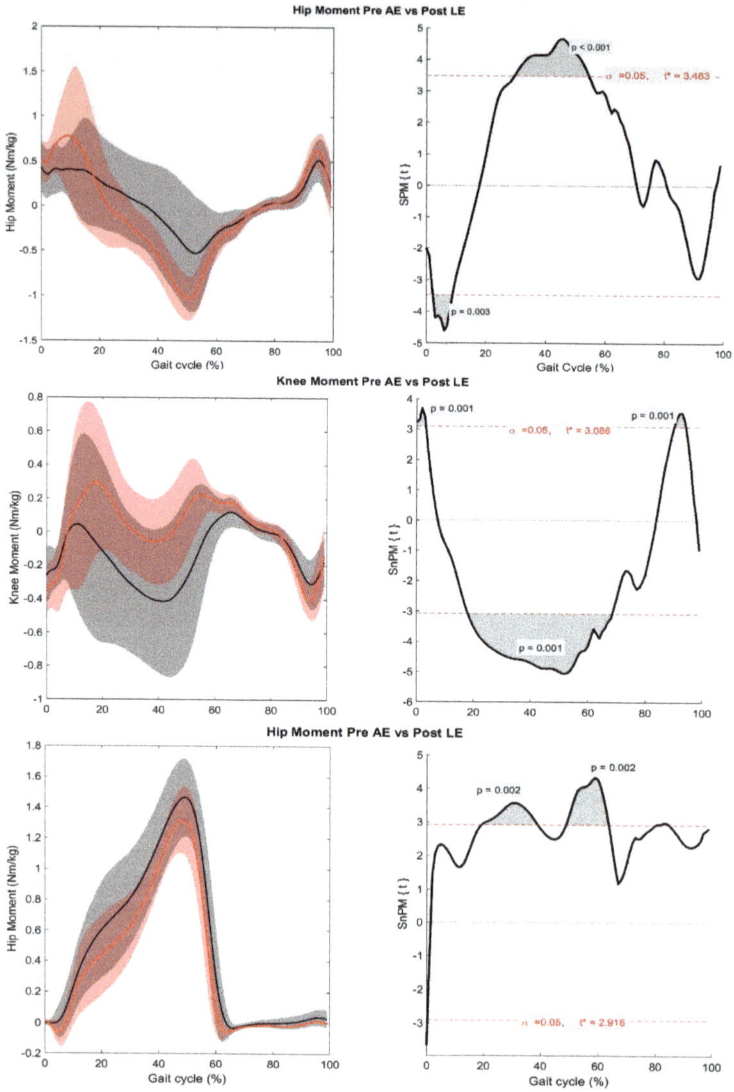

Figure 4. Hip, knee, and ankle joint moments in Pre AE and Post LE assessments. In the three left panels, the black lines correspond to the average of Pre AE trajectories, while the red lines represent Post LE; shaded areas indicate the standard deviation. Positive moment values indicate extension and plantarflexion moments, while negative values denote flexion and dorsiflexion moments. In the three right panels, the black lines correspond to the SPM or SnPM calculated t-value, and t* is the critical threshold. Grey areas above or below the red dotted lines indicate a significant difference ($p < 0.008$). Abbreviations: Pre AE, pre-aquatic exercise assessment; Post LE, post-land exercise assessment.

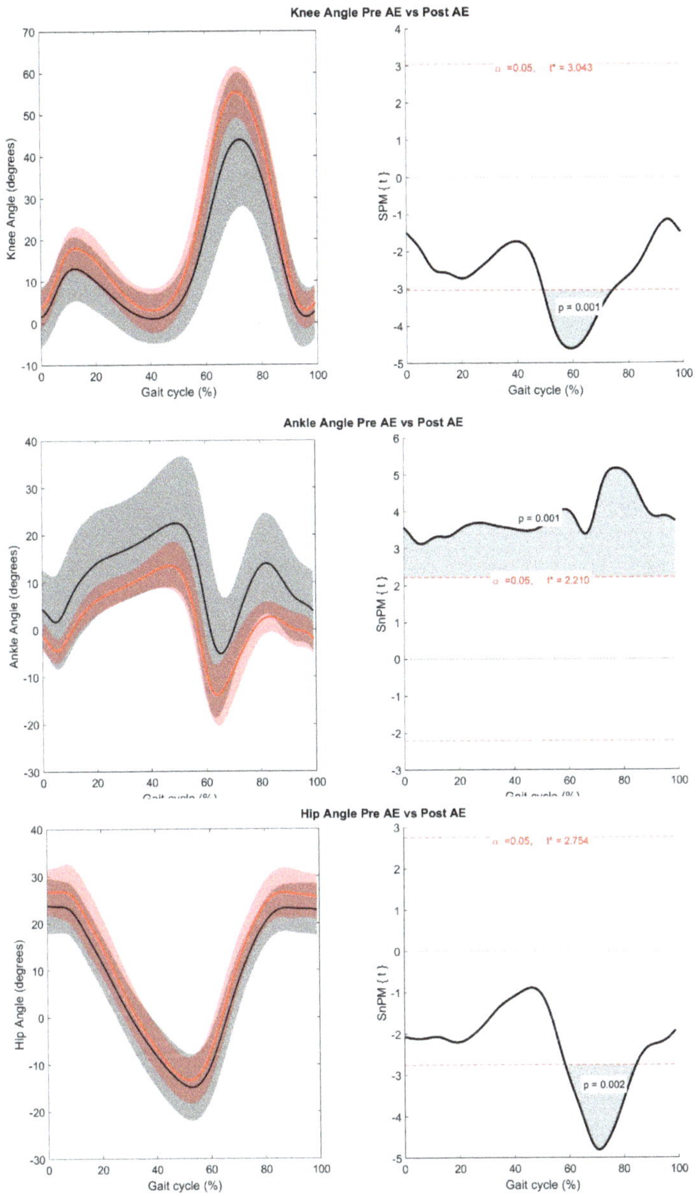

Figure 5. Hip, knee, and ankle joint angles in Pre AE and Post LE assessments. In the three left panels, the black lines correspond to the average of Pre AE trajectories, while the red lines correspond to Post LE; shaded areas represent the standard deviation; positive angle values indicate flexion and dorsiflexion, whereas negative values represent extension and plantarflexion angles. In the three right panels, black lines represent the SPM or SnPM calculated t-value, and t* is the critical threshold. Grey areas above or below the red dotted lines indicate a significant difference ($p < 0.008$). Abbreviations: Pre AE, pre-aquatic exercise assessment; Post LE, post-land exercise assessment.

4. Discussion

In this study, we investigated biomechanical changes in the gait of individuals with Parkinson's disease (PD) over nine months while they participated in a multimodal exercise program in both aquatic and land environments. Following training in both environments, we observed changes in spatiotemporal, kinematic, and kinetic gait parameters.

A spatiotemporal reorganization of gait was evident, characterized by a decrease in double support time and an increase in cadence (see Table 1). There was a significant increase in gait speed following land-based exercise (post-LE) with an effect size classified as medium. Also, the observed change is larger than the moderate clinically meaningful difference in gait speed among individuals with PD in the on-medication state, which is 0.14 m/s [41]. The observed reduction in double support time might indicate enhanced postural control, as the center of mass remains within the base of support primarily during the double support phase of the gait cycle [42]. Notably, the decrease in stride length did not compensate for the increase in cadence. This observation is significant, given that a reduction in step length associated with increased cadence is characteristic of natural PD progression [43].

Gait speed consistently increased throughout the study, even during the retention period, which is unexpected in a neurodegenerative disease like PD. This may be attributed to the development of new motor strategies prompted by aquatic exercises. In fact, Volpe et al. [6] found differences in muscle activation among PD participants even several days after discontinuing underwater training activities. Moreover, recent research suggests that shared muscle synergies across different motor tasks maximize the generalization of motor learning effects [24]. Thus, our results hint at the transfer of skills from water to land environments. Future studies could examine this issue by analyzing kinematic, kinetic, or muscle activity patterns and identifying synergies.

Joint moments change significantly from pre-AE to post-LE (Figure 4). The knee flexor moment increases, associated with the eccentric contraction of the knee extensors to control weight acceptance [44]. The ankle moment shifts towards a dorsiflexor curve, indicating improved plantar flexion control after heel contact and better weight absorption in post-LE. The increased hip and knee extensor moments during the loading response relate to anterior trunk acceleration control due to the rapid body weight transfer onto the foot [45].

Due to the action of bi-articular muscles, joint moments influence adjacent segments. Therefore, changes in the hip moment are expected to occur concurrently with changes in the knee and ankle moments, as observed here. Considering the bi-articular nature of the rectus femoris, which spans both the hip and knee, coordinated improvements in hip and knee extensor moments—previously described by Winter [46] as part of a synergistic pattern—are expected. Such changes can also be influenced by improved motor unit recruitment, reduced co-contraction, and more efficient timing of muscle activation, all of which contribute to increased net joint torque production. Therefore, the observed improvements in gait cadence and velocity are likely the result of integrated neuromuscular and biomechanical enhancements, rather than isolated strength gains. However, as we did not use any other form of assessment, such as electromyography, we cannot categorically state that this was the mechanism underlying the observation.

During the mid and terminal stance phases, the hip flexor moment gradually increased, peaking at approximately 50% of the gait cycle (see Figure 4). In pre-AE, the knee exhibited a flexor moment throughout the stance phase. However, in post-LE, this pattern shifted towards a trajectory like that of healthy individuals, characterized by a double extension curve, with the peak of the extensor moment occurring at mid-stance [44]. This increase in hip flexion and knee extension moments may suggest improved alignment of the ground reaction vector, facilitating lower limb extension. In individuals with PD, this alignment may be associated with reduced energy consumption, as improved positioning and higher moments can help spare muscles from generating unnecessary contractions. To validate these hypotheses, future studies should analyze the ground reaction forces.

The ankle exhibited a reduced plantar flexor moment throughout the stance phase, including its peak value during the terminal stance/pre-swing post-LE (see Figure 4). This appears unexpected at first glance because the increase in velocity requires a larger impulse at foot-off. Our explanation is that subjects increased the knee extension moment to enhance the impulse without increasing the ankle plantarflexion moment. It is well-documented that the ankle is the most affected joint by kinematic and kinetic deficits in PD [10] and that individuals with PD generate less ankle power during the terminal stance phase than their healthy counterparts [9]. Moreover, Skinner et al. [47] reported that people with PD have a diminished capacity to produce ankle joint moments, forcing the adoption of alternative control strategies. We believe this is what occurred with the individuals in this study.

This compensatory mechanism aligns with previous findings in the literature. For instance, a recent treadmill-based study in individuals with PD demonstrated that, although dopaminergic medication increased gait speed and propulsion, improvements in joint torque occurred primarily at the hip, while ankle plantarflexion moments remained unchanged [48]. Similarly, Albani et al. [9] reported reduced ankle power in individuals with PD walking at their preferred speed, suggesting that distal musculature contributes less to propulsion under dopaminergic deficits. These observations support the hypothesis that, when ankle push-off capacity is compromised, individuals with PD increasingly rely on proximal joints—particularly the hip and knee—to maintain gait velocity. A similar redistribution of joint kinetics has been reported in healthy older adults [8,49]. Those authors emphasize that increased torque generation at proximal joints may relieve the ankle of its mechanical demands. Taken together, this body of evidence reinforces the interpretation that the reduction in ankle plantarflexor moment observed in this study—despite increased walking speed—reflects a proximal compensation strategy rather than a contradiction of expected gait mechanics.

In the terminal swing phase, a significantly higher knee flexor moment is observed post-LE (see Figure 4). This increase leads to typical values, as reported by Sloot et al. [44], and may be associated with the eccentric contraction of knee flexors, which decelerates the knee and prepares for ground contact [45].

Regarding the kinematics (Figure 5), significant increases in hip and knee peak flexion are observed post-LE during the initial swing phase. The coordinated movement of these joints facilitates the swing phase, enhancing foot clearance. Significantly, this increase in peak flexion during the initial swing phase does not interfere with the extension of either joint. This observation is significant given the tendency in PD to exhibit increased lower limb and trunk flexor patterns [50]. Furthermore, the ankle angle trajectory has shifted towards lower values throughout the entire stride, coming closer to normal gait. It is well-known that PD gait typically shows increased ankle dorsiflexion compared to normal [9,51]. After nine months, dorsiflexion at foot strike was absent. Across the entire stride, dorsiflexion was reduced, while plantarflexion was increased.

It is important to emphasize that this is an interventional cohort study, without a control group, which is a significant limitation. As such, it is not possible to attribute the observed changes solely to the exercise program. Results could be attributable to confounding factors, such as the natural progression of the disease or placebo effects. However, it is worth noticing that our results contrast with the expected decline in spatiotemporal variables as Parkinson's disease progresses. It is well known that, over the years, Parkinson's disease leads to a reduction in step length, with an increase in double support phase and cadence, resulting in shorter and more frequent step exchanges [1,43].

A key limitation of this study is its small sample size, especially due to the loss of five participants. As a result, the statistical power is limited. Additionally, the dropout of participants might have introduced bias, since those who remained could be more motivated or more concerned about their health.

Additionally, they were all recruited from the same Parkinson's Association, which might have introduced some bias, as those who participate in such institutions could be more motivated and in better overall health than those who do not. We recognize that these factors reduce the external validity of the study. On the other hand, participants' involvement in the same PD association facilitated consistent communication and monitoring, which likely supported adherence to the study guidelines.

Despite the study design and sample limitations, we consider our results valuable, given the burden of PD on healthcare systems. Families and patients also suffer from the hopelessness of a degenerative disease. The exercise protocols are sound and based on recognized international guidelines. Therefore, we recommend that future studies with more robust designs, such as multicentric randomized clinical trials or action observation therapy with one or both exercise protocols, be conducted. The challenges of standardizing the therapy across centers could be addressed using tele-rehabilitation strategies [52]. Moreover, the biomechanical evaluation could be performed in clinical settings using markerless motion capture systems [53].

Regarding biomechanical data, we have not analyzed ground reaction forces (GRFs) in this study. Although we recognize its importance, we believe it would not add much information, as our focus was on joint moments. Since the moments were calculated using Nexus software, we do not have the exact equations from it. Thus, establishing a mechanistic relationship between GRFs and the moments would not be possible anyway. Future studies could address this limitation through biomechanical and musculoskeletal modeling [53,54]. Incorporating musculoskeletal modeling would help in understanding the underlying mechanisms of the observed biomechanical changes and provide insights for developing motion strategies such as those in [54].

In summary, despite this study's limitations, our findings collectively suggest beneficial changes in gait biomechanics, as hypothesized. Thus, encouraging future studies with more robust designs, adding other measures such as EMG, using markerless motion tracking, and incorporating musculoskeletal modeling.

5. Conclusions

Changes in spatiotemporal parameters, kinetics, and gait kinematics occurred following the nine months of intervention. Walking speed increased, while the duration of the double support phase decreased. Additionally, the knee extensor moment consistently increased throughout the interval from midstance to midswing (20% to 70% of the stride period), approaching standard gait patterns. Significant increases were also observed in both hip and knee peak flexion angles. Furthermore, the abnormal ankle dorsiflexion observed at foot strike disappeared.

Author Contributions: Conceptualization, D.D.I. and V.L.I.; data curation, D.D.I.; formal analysis, E.F.M.; funding acquisition, D.D.I. and V.L.I.; investigation, D.D.I.; methodology, D.D.I., E.F.M. and V.L.I.; project administration, V.L.I.; resources, D.D.I. and V.L.I.; software, E.F.M.; supervision, E.F.M. and V.L.I.; validation, D.D.I., E.F.M. and V.L.I.; visualization, E.F.M.; writing—original draft, D.D.I.; writing—review and editing, D.D.I. and E.F.M. All authors have read and agreed to the published version of the manuscript.

Funding: This study was financed in part by the Coordenação de Aperfeiçoamento de Pessoal de Nível Superior—Brasil (CAPES)—Finance Code 001.

Institutional Review Board Statement: This study was conducted in accordance with the Declaration of Helsinki and approved by the Federal University of Paraná Human Research Ethics Committee (protocol 3.780.224) for studies involving humans.

Informed Consent Statement: Informed consent was obtained from all subjects involved in this study.

Data Availability Statement: According to the Research Ethics Committee of the Federal University of Paraná (UFPR), Brazil, research data cannot be shared due to ethical issues regarding their secrecy and nondisclosure. Only the study results may be shared.

Acknowledgments: The authors would like to thank Luiz Augusto Kalva for his help with the scripts for data processing.

Conflicts of Interest: The authors declare no conflicts of interest.

Abbreviations

The following abbreviations are used in this manuscript:

PD	Parkinson's Disease
SPM	Statistical Parametric Mapping
AE	Aquatic Exercise
H&Y	Modified Hoehn and Yahr
UPDRS	Unified Parkinson's Disease Rating Scale
LE	Land Exercise

References

1. Kalia, L.V.; Lang, A.E. Parkinson's disease. *Lancet* **2015**, *386*, 896–912. [CrossRef]
2. Titova, N.; Padmakumar, C.; Lewis, S.J.G.; Chaudhuri, K.R. Parkinson's: A syndrome rather than a disease? *J. Neural Transm.* **2017**, *124*, 907–914. [CrossRef]
3. Mirelman, A.; Bonato, P.; Camicioli, R.; Ellis, T.D.; Giladi, N.; Hamilton, J.L.; Hass, C.J.; Hausdorff, J.M.; Pelosin, E.; Almeida, Q.J. Gait impairments in Parkinson's disease. *Lancet Neurol.* **2019**, *18*, 697–708. [CrossRef] [PubMed]
4. Zanardi, A.P.J.; Silva, E.S.; Costa, R.R.; Passos-Monteiro, E.; Santos, I.O.; Kruel, L.F.M.; Peyré-Tartaruga, L.A. Gait parameters of Parkinson's disease compared with healthy controls: A systematic review and meta-analysis. *Sci. Rep.* **2021**, *11*, 752. [CrossRef]
5. Spolaor, F.; Volpe, D.; Pavan, D.; Guiotto, A.; Fichera, F.; Torresin, P.; Fantinato, E.; Sawacha, Z. Surface EMG analysis in Parkinson disease patients before and after underwater gait training. *Gait Posture* **2017**, *57*, 185–191. [CrossRef]
6. Volpe, D.; Spolaor, F.; Sawacha, Z.; Guiotto, A.; Pavan, D.; Bakdounes, L.; Urbani, V.; Frazzitta, G.; Iansek, R. Muscular activation changes in lower limbs after underwater gait training in Parkinson's disease: A surface EMG pilot study. *Gait Posture* **2020**, *80*, 185–191. [CrossRef]
7. Morris, M.E.; McGinley, J.; Huxham, F.; Collier, J.; Iansek, R. Constraints on the kinetic, kinematic and spatiotemporal parameters of gait in Parkinson's disease. *Hum. Mov. Sci.* **1999**, *18*, 461–483. [CrossRef]
8. Kuhman, D.; Hammond, K.G.; Hurt, C.P. Altered joint kinetic strategies of healthy older adults and individuals with Parkinson's disease to walk at faster speeds. *J. Biomech.* **2018**, *79*, 112–118. [CrossRef] [PubMed]
9. Albani, G.; Cimolin, V.; Fasano, A.; Trotti, C.; Galli, M.; Mauro, A. "Masters and servants" in parkinsonian gait: A three-dimensional analysis of biomechanical changes sensitive to disease progression. *Funct. Neurol.* **2014**, *29*, 99–105.
10. Sofuwa, O.; Nieuwboer, A.; Desloovere, K.; Willems, A.-M.; Chavret, F.; Jonkers, I. Quantitative gait analysis in Parkinson's disease: Comparison with a healthy control group. *Arch. Phys. Med. Rehabil.* **2005**, *86*, 1007–1013. [CrossRef]
11. Ernst, M.; Folkerts, A.K.; Gollan, R.; Lieker, E.; Caro-Valenzuela, J.; Adams, A.; Cryns, N.; Monsef, I.; Dresen, A.; Roheger, M.; et al. Physical exercise for people with Parkinson's disease: A systematic review and network meta-analysis. *Cochrane Database Syst. Rev.* **2023**, *1*, CD013856. [CrossRef] [PubMed]
12. Ni, M.; Hazzard, J.B.; Signorile, J.F.; Luca, C. Exercise Guidelines for Gait Function in Parkinson's Disease: A Systematic Review and Meta-analysis. *Neurorehabil. Neural Repair* **2018**, *32*, 872–886. [CrossRef]
13. Hvingelby, V.S.; Glud, A.N.; Sørensen, J.C.H.; Tai, Y.; Andersen, A.S.M.; Johnsen, E.; Moro, E.; Pavese, N. Interventions to improve gait in Parkinson's disease: A systematic review of randomized controlled trials and network meta-analysis. *J. Neurol.* **2022**, *269*, 4068–4079. [CrossRef]

14. Keus, S.H.; Munneke, M.; Graziano, M.; Paltamaa, J.; Pelosin, E.; Domingos, J.; Bloem, B. *European Physiotherapy Guideline for Parkinson's Disease*; KNGF/ParkinsonNet: Nijmegen, The Netherlands, 2014.
15. Osborne, J.A.; Botkin, R.; Colon-Semenza, C.; DeAngelis, T.R.; Gallardo, O.G.; Kosakowski, H.; Martello, J.; Pradhan, S.; Rafferty, M.; Readinger, J.L.; et al. Physical therapist management of Parkinson disease: A clinical practice guideline from the American Physical Therapy Association. *Phys. Ther.* **2022**, *102*, pzab302. [CrossRef]
16. Rafferty, M.R.; Prodoehl, J.; Robichaud, J.A.; David, F.J.; Poon, C.; Goelz, L.C.; Vaillancourt, D.E.; Kohrt, W.M.; Comella, C.L.; Corcos, D.M. Effects of two years of exercise on gait impairment in people with Parkinson's Disease: The PRET-PD randomized trial. *J. Neurol. Phys. Ther.* **2017**, *41*, 21–30. [CrossRef]
17. Shen, X.; Mak, M.K. Balance and gait training with augmented feedback improves balance confidence in people with Parkinson's disease: A randomized controlled trial. *Neurorehabil. Neural Repair* **2014**, *28*, 524–535. [CrossRef]
18. Pinto, C.; Salazar, A.P.; Marchese, R.R.; Spector, N.D.; Mukherjee, D. The effects of hydrotherapy on balance, functional mobility, motor status, and quality of life in patients with Parkinson disease: A systematic review and meta-analysis. *PMR* **2019**, *11*, 278–291. [CrossRef]
19. Gomes Neto, M.; Pontes, S.S.; Almeida, L.D.; da Silva, C.M.; Sena, C.; Saquetto, M.B. Effects of water-based exercise on functioning and quality of life in people with Parkinson's disease: A systematic review and meta-analysis. *Clin. Rehabil.* **2020**, *34*, 1425–1435. [CrossRef] [PubMed]
20. Volpe, D.; Giantin, M.G.; Maestri, R.; Frazzitta, G. Comparing the effects of hydrotherapy and land-based therapy on balance in patients with Parkinson's disease: A randomized controlled pilot study. *Clin. Rehabil.* **2014**, *8*, 1210–1217. [CrossRef]
21. Volpe, D.; Pavan, D.; Morris, M.; Guiotto, A.; Iansek, R.; Fortuna, S.; Frazzitta, G.; Sawacha, Z. Underwater gait analysis in Parkinson's disease. *Gait Posture* **2017**, *52*, 87–94. [CrossRef] [PubMed]
22. Orselli, M.I.; Duarte, M. Joint forces and torques when walking in shallow water. *J. Biomech.* **2011**, *44*, 1170–1175. [CrossRef] [PubMed]
23. Miyoshi, T.; Shirota, T.; Yamamoto, S.I.; Nakazawa, K.; Akai, M. Functional roles of lower-limb joint moments while walking in water. *Clin. Biomech.* **2005**, *20*, 194–201. [CrossRef] [PubMed]
24. Yokoyama, H.; Kato, T.; Kaneko, N.; Kobayashi, H.; Hoshino, M.; Kokubun, T.; Nakazawa, K. Basic locomotor muscle synergies used in land walking are finely tuned during underwater walking. *Sci. Rep.* **2021**, *11*, 18480. [CrossRef] [PubMed]
25. Bakhtiyarian, R.; Majlesi, M.; Azadian, E.; Ali, M.J. Examining virtual reality's influence on kinetic variables for obstacle crossing in Parkinson's disease. *Gait Posture* **2025**, *121*, 85–92. [CrossRef]
26. Russo, M.; Amboni, M.; Pisani, N.; Volzone, A.; Calderone, D.; Barone, P.; Amato, F.; Ricciardi, C.; Romano, M. Biomechanics Parameters of Gait Analysis to Characterize Parkinson's Disease: A Scoping Review. *Sensors* **2025**, *25*, 338. [CrossRef]
27. Pataky, T.C. Generalized n-dimensional biomechanical field analysis using statistical parametric mapping. *J. Biomech.* **2010**, *43*, 1976–1982. [CrossRef]
28. Goetz, C.G.; Fahn, S.; Martinez-Martin, P.; Poewe, W.; Sampaio, C.; Stebbins, G.T.; Stern, M.B.; Tilley, B.C.; Dodel, R.; Dubois, B.; et al. Movement Disorder Society-sponsored revision of the Unified Parkinson's Disease Rating Scale (MDS-UPDRS): Process, format, and clinimetric testing plan. *Mov. Disord.* **2007**, *22*, 41–47. [CrossRef]
29. Schenkman, M.L.; Clark, K.; Xie, T.; Kuchibhatla, M.; Shinberg, M.; Ray, L. Spinal movement and performance of a standing reach task in participants with and without Parkinson disease. *Phys. Ther.* **2001**, *81*, 1400–1411. [CrossRef]
30. Davis, R.B., III; Ounpuu, S.; Tyburski, D.; Gage, J.R. A gait analysis data collection and reduction technique. *Hum. Mov. Sci.* **1991**, *10*, 575–587. [CrossRef]
31. Carroll, L.M.; Morris, M.E.; O'Connor, W.T.; Volpe, D.; Salsberg, J.; Clifford, A.M. Evidence-Based Aquatic Therapy Guidelines for Parkinson's Disease: An International Consensus Study. *J. Parkinsons Dis.* **2022**, *12*, 621–637. [CrossRef]
32. Siega, J.; Iucksch, D.D.; Israel, V.L. Multicomponent Aquatic Training (MAT) Program for People with Parkinson's Disease: A Protocol for a Controlled Study. *Int. J. Environ. Res. Public Health* **2022**, *19*, 1727. [CrossRef]
33. Iucksch, D.D. Análise dos Efeitos de um Programa de Exercícios Físicos Multicomponentes em Ambiente Aquático E Terrestre na Funcionalidade, na Função Muscular, no Controle Postural e na Marcha de Pessoas Com Doença de Parkinson. Ph.D. Thesis, Federal University of Paraná, Curitiba, Brazil, 2022.
34. Kanitz, A.C.; Delevatti, R.S.; Reichert, T.; Liedtke, G.V.; Ferrari, R.; Almada, B.P.; Pinto, S.S.; Alberton, C.L.; Kruel, L.F.M. Effects of two deep water training programs on cardiorespiratory and muscular strength responses in older adults. *Exp. Gerontol.* **2015**, *64*, 55–61. [CrossRef]
35. Thompson, P.D.; Arena, R.; Riebe, D.; Pescatello, L.S.; American College of Sports Medicine. ACSM's new preparticipation health screening recommendations from ACSM's guidelines for exercise testing and prescription, ninth edition. *Curr. Sports Med. Rep.* **2013**, *12*, 215–217. [CrossRef] [PubMed]
36. Lima, L.O.; Scianni, A.; Rodrigues-De-Paula, F. Progressive resistance exercise improves strength and physical performance in people with mild to moderate Parkinson's disease: A systematic review. *J. Physiother.* **2013**, *59*, 7–13. [CrossRef] [PubMed]

37. Cunha, M.C.B.; Alonso, A.C.; Silva, T.M.; de Raphael, A.C.B.; Mota, C.F. Ai Chi: Efeitos do relaxamento aquático no desempenho funcional e qualidade de vida em idosos. *Fisioter. Em Mov.* **2010**, *23*, 409–417. [CrossRef]
38. Pereira, M.P.; Silva, N.A.; Matida, A.B.; Costa, J.N.A.; Gonçalves, C.D.; Safons, M.P.; Vianna, L.G. Protocolo de intervenção de Tai Chi Chuan para idosos. *Lect. Educ. Física Y Deportes* **2009**, *14*, 139.
39. Cohen, J. *Statistical Power Analysis for the Behavioral Sciences*; Lawrence Erlbaum Associates: Hillsdale, NJ, USA, 1988.
40. Donnelly, C.J.; Alexander, C.; Pataky, T.C.; Stannage, K.; Reid, S.; Robinson, M. Vector-field statistics for the analysis of time varying clinical gait data. *Clin. Biomech.* **2017**, *41*, 87–91. [CrossRef]
41. Hass, C.J.; Bishop, M.; Moscovich, M.; Stegemöller, E.L.; Skinner, J.; Malaty, I.A.; Shukla, A.W.; McFarland, N.; Okun, M.S. Defining the clinically meaningful difference in gait speed in persons with Parkinson disease. *J. Neurol. Phys. Ther.* **2014**, *38*, 233–238. [CrossRef]
42. Winter, D.A. Human balance and posture control during standing and walking. *Gait Posture* **1995**, *3*, 193–214. [CrossRef]
43. Fasano, A.; Bloem, B.R. Gait disorders. *Continuum* **2013**, *19*, 1344–1382. [CrossRef]
44. Sloot, L.H.; van der Krogt, M.M. Interpreting Joint Moments and Powers in Gait. In *Handbook of Human Motion*; Müller, B., Wolf, S.I., Brueggemann, G.P., Deng, Z., McIntosh, A.S., Miller, F., Selbie, W.S., Eds.; Springer: Cham, Switzerland, 2018; pp. 625–643. [CrossRef]
45. Winter, D.A. *Biomechanics and Motor Control of Human Movement*, 4th ed.; John Wiley & Sons: Hoboken, NJ, USA, 2009. [CrossRef]
46. Winter, D.A. Biomechanics of normal and pathological gait: Implications for understanding human locomotor control. *J. Mot. Behav.* **1989**, *21*, 337–355. [CrossRef]
47. Skinner, J.W.; Lee, H.K.; Roemmich, R.T.; Amano, S.; Hass, C.J. Execution of activities of daily living in persons with Parkinson disease. *Med. Sci. Sports Exerc.* **2015**, *47*, 1906–1912. [CrossRef]
48. Baudendistel, S.T.; Schmitt, A.C.; Roemmich, R.T.; Harrison, I.L.; Hass, C.J. Levodopa facilitates improvements in gait kinetics at the hip, not the ankle, in individuals with Parkinson's disease. *J. Biomech.* **2021**, *121*, 110366. [CrossRef]
49. Devita, P.; Hortobagyi, T. Age causes a redistribution of joint torques and powers during gait. *J. Appl. Physiol.* **2000**, *88*, 1804–1811. [CrossRef] [PubMed]
50. Morris, M.E.; Huxham, F.; McGinley, J.; Dodd, K.; Iansek, R. The biomechanics and motor control of gait in Parkinson disease. *Clin. Biomech.* **2001**, *16*, 459–470. [CrossRef] [PubMed]
51. Pistacchi, M.; Gioulis, M.; Sanson, F.; De Giovannini, E.; Filippi, G.; Rossetto, F.; Marsala, S.Z. Gait analysis and clinical correlations in early Parkinson's disease. *Funct. Neurol.* **2017**, *32*, 28–34. [CrossRef] [PubMed]
52. Truijen, S.; Abdullahi, A.; Bijsterbosch, D.; van Zoest, E.; Conijn, M.; Wang, Y.; Struyf, N.; Saeys, W. Effect of home-based virtual reality training and telerehabilitation on balance in individuals with Parkinson disease, multiple sclerosis, and stroke: A systematic review and meta-analysis. *Neurol. Sci.* **2022**, *43*, 2995–3006. [CrossRef] [PubMed] [PubMed Central]
53. Uhlrich, S.D.; Falisse, A.; Kidziński, Ł.; Muccini, J.; Ko, M.; Chaudhari, A.S.; Hicks, J.L.; Delp, S.L.; Marsden, A.L. OpenCap: Human movement dynamics from smartphone videos. *PLoS Comput. Biol.* **2023**, *19*, e1011462. [CrossRef]
54. Xu, D.; Zhou, H.; Quan, W.; Ma, X.; Chon, T.-E.; Fernandez, J.; Gusztav, F.; Kovács, A.; Baker, J.S.; Gu, Y. New Insights Optimize Landing Strategies to Reduce Lower Limb Injury Risk. *Cyborg Bionic Syst.* **2024**, *5*, 0126. [CrossRef]

biomechanics

MDPI

Article

Evaluation of Voluntary Dynamic Balance through Standardized Squat-Lift Movements: A Comparison between Gymnasts and Athletes from Other Sports

Jair Wesley Ferreira Bueno [1,*], Daniel Boari Coelho [1,2] and Luis Augusto Teixeira [1]

1 Human Motor Systems Laboratory, School of Physical Education and Sport, University of São Paulo, USP, Av. Prof. Mello Moraes, 65. Cidade Universitária, São Paulo 05508-030, SP, Brazil; daniel.boari@ufabc.edu.br (D.B.C.); lateixei@usp.br (L.A.T.)

2 Biomedical Engineering, Federal University of ABC, São Bernardo do Campo 09606-070, SP, Brazil

* Correspondence: jwesleybueno@usp.br; Tel.: +55-35-99109-6005

Abstract: In the quotidian, people perform voluntary whole-body movements requiring dynamic body balance. However, the literature is scarce of dynamic balance evaluations employing standardized voluntary movements. In this investigation, we aimed to analyze the sensitivity of balance evaluation between gymnasts and athletes from other sports in the performance of balance tasks. Participants were evaluated in upright quiet standing and the performance of cyclic dynamic tasks of hip flexion-extension and squat-lift movements. Movements were individually standardized in amplitude, while the rhythm was externally paced at the frequency of 0.5 Hz. Tasks were performed on a force plate, with dynamic balance measured through the center of pressure displacement. Results showed that in quiet standing and the dynamic hip flexion-extension task, no significant differences were found between the groups. Conversely, results for the squat-lift task revealed a better balance of the gymnasts over controls, as indicated by the reduced amplitude and velocity of the center of pressure displacement during the task execution. The superior balance performance of gymnasts in the squat-lift task was also observed when vision was suppressed. These findings suggest the employed squat-lift task protocol is a potentially sensitive procedure for the evaluation of voluntary dynamic balance.

Keywords: equilibrium; evaluation; young; center of pressure; protocol

1. Introduction

Much of the current knowledge on balance control has been acquired through the assessment of quiet standing, characterized by keeping an upright motionless stance. On the other hand, our daily living activities are characterized by dynamic balance, with the maintenance of stance while performing voluntary movements with the trunk and limbs, like standing up from a chair or in manual reaching. Investigation of dynamic balance has recently attracted scientific interest ([1], for a review), with research aiming at developing reliable and valid evaluation protocols (cf. [2]). In functional or clinical assessment, Y-balance [3], star excursion [4], and timed up-and-go [5] tests have been employed as proxy measurements of dynamic balance. Performance on the Y-balance and star excursion tests is measured through the maximum distance one can move a single foot in different directions over the ground in unipedal stance. Beyond requiring dynamic balance, performance on these tests has been shown to be affected by joints' range of motion [6,7] and strength of hip extensor muscles [4]. The timed up-and-go test is evaluated through the completion time to stand up from a chair, walk 3 m straightforwardly, return to the chair, and sit down. Performance on this test is mainly affected by the legs' muscular power [8]. Thus, as these tests are affected by different confounding factors, the respective measurements based on range of motion or completion time could not be taken as faithful indexes of dynamic balance.

An alternative to achieve accurate and valid measurements of dynamic balance is employing biomechanical assessments of body stability when moving, as indexed by different variables related to the center of mass or the center of pressure displacement. The prevalent research strategy for a biomechanical analysis of dynamic balance has been assessing reactive responses to intrinsic or extrinsic perturbations to stance. Some variations of this experimental strategy include the following: (a) reacting to unexpected translations [9] or rotations [10] of the support base, (b) recovering balance after the release of a load attached to the trunk leading to fast body sway [11], balancing on (c) a continuously moving platform [12] or (d) on an unstable support board [13]. Whereas objective biomechanical measurements have provided valid and reliable assessments in these reactive balance tasks (cf. [9]), results are task-specific, with a lack of association with balance measurements either in quiet standing [9,14] or with voluntary movements [15]. In the study by Ringhof and Stein [15], in particular, gymnasts and swimmers were compared on three balance tasks, requiring balance recovery from self-perturbations induced by an unstable support base or by a mechanically provoked fast forward body sway, in addition to a voluntary task of one-leg landing after short horizontal jumping. Results showed that the expected higher balance performance of the gymnasts over swimmers was found only in the voluntary landing task, while in the reactive balance tasks, performance was found to be equivalent between the two groups. From these findings, it seems that task-specific measurements are required for an accurate evaluation of voluntary dynamic balance.

One of the experimental strategies employed to evaluate the effect of prior experience on balance control involves comparing athletes from different sports. These athletes are exposed to varying balance demands during their routine training sessions. A literature review has shown that athletes have greater balance stability than non-athletes and high-level athletes have better balance control than low-level athletes [16]. Gymnasts (gymnastics is a type of sport that involves physical exercises requiring balance, strength, flexibility, agility, coordination, artistry, and endurance. Gymnasts often perform controlled movements on special equipment, such as bars, beams, and mats) and, in particular, have been found to develop high balance skills compared with athletes from several other sports [17]. Further research has supported the notion that gymnasts have increased balance in comparison with individuals regularly exposed to less demanding balance tasks. For instance, Davlin [18] compared high-level gymnasts, soccer players, and swimmers, having non-athletes as controls, in a dynamic balance task of standing on a stabilometer. Results revealed that gymnasts had higher balance stability than all other groups. Gymnasts and experts in other sports were compared in balance tasks with different difficulty levels, ranging from full vision in a bipedal stance on a rigid surface to unipedal standing on a malleable surface with visual occlusion [19]. Results indicated that gymnasts had higher balance stability in the more challenging balance tasks represented by no vision and distorted somesthetic information from the feet soles due to the malleable surface. This result suggests that the increased balance proficiency of gymnasts can be detected in more challenging tasks, like those involving sensory manipulation. Vuillerme and Nougier [20] assessed attentional demands between expert gymnasts and expert performers in other non-gymnastic sports in the manual task of responding as quickly as possible to an unpredictable auditory stimulus. The manual task was performed while standing with different balance demands, including manipulation of area and malleability of the support base. Results revealed lower attentional demands in the gymnasts than in controls during the challenging unipedal stance. This finding can be interpreted as indicating increased gymnasts' automaticity in regulating the required anticipatory balance adjustments to prevent stance perturbations potentially induced by the voluntary movement (for further evidence of improved balance control in gymnasts, see [21–23]). From the reviewed results, one can assume that gymnasts represent an appropriate reference for testing probing tasks for the evaluation of voluntary dynamic balance.

A critical point for the appropriate evaluation of voluntary dynamic balance is setting test constraints to achieve similar movements across individuals during the evaluation. Movements performed with different amplitudes or rhythms can affect objective measurements of balance stability, imposing difficulties in the interpretation of balance control. A preliminary attempt to standardize voluntary movements for the evaluation of dynamic balance was made by Bueno et al. [24]. The task consisted of performing cyclic hip flexions and extensions, so that the hip was flexed at about 45 degrees at the extreme position, assuming then the upright posture at the end of the cycle. Movements were standardized in amplitude, while the rhythm of the repeated movements was paced through beeps emitted by a metronome at regular intervals. Another potential task for the evaluation of dynamic balance is the cyclic sit-to-stand task. Research has shown that the completion time to perform the functional five times sit-to-stand test [25] is importantly affected by dynamic balance [26–29]. From these findings, both cyclic hip flexion extension and sit-to-stand movements can be conceived to be potential candidates for a reliable assessment of voluntary dynamic balance. In the current investigation, we had as the primary aim to test the sensibility of tasks requiring cyclic hip flexion-extension and squat-lift (similar to sit-to-stand movements) for assessment of voluntary dynamic balance, comparing groups of gymnasts and non-gymnasts. The underlying rationale for this comparison is that if the tasks provide a sensitive and reliable evaluation of voluntary dynamic balance, the expected increased balance control of gymnasts should be reflected in objective measurements of body stability. As the secondary aim, we tested the extent to which visual occlusion affects voluntary dynamic balance.

2. Materials and Methods

2.1. Participants

Male athletes from gymnastics (n = 9) and from other sports (n = 10) participated in this study. The gymnasts were high-level athletes at the adult national level, with 3 of them making part of the national team. They trained in the sport for at least 5 consecutive years, with the most experienced athlete accumulating 20 years of training. At the time of testing, they were training with an average frequency of 6 times per week, completing 24–48 h of training per week across participants. The comparison group was composed of athletes from different sports, as follows: soccer (n = 3), rugby (n = 3), squash (n = 1), basketball (n = 2), and athletics (n = 1). Participants of this comparison group had a minimum of 5 years of practice in the trained sport, training with frequencies of 3–5 times per week. Table 1 shows participants' descriptive information separately for each group. In addition to expertise in the trained sport, the inclusion criteria were the absence of lower limb injuries at the time of testing; no participants were excluded. A post hoc estimation of the power of the sample size was made through G*Power (Heinrich-Heine-Universität Düsseldorf, Düsseldorf, Germany; http://www.gpower.hhu.de/, (accessed on 17 May 2024)) for repeated measures, within-between group by vision interaction, effect size = 0.25, α = 0.05. The result indicated a power of 0.70 for our sample size.

Table 1. Age, anthropometric data, and training times separately by group.

	Gymnasts	Other Athletes
Age (years)	20.44 (4.33)	22.7 (2.67)
Weight (Kg)	64.26 (7.41)	80.10 (12.38)
Height (cm)	167.67 (4.36)	178.80 (7.19)
Weekly training (h)	32.11 (5.27)	6.10 (5.57)
Weekly frequency (days)	6.00 (0)	4.20 (1.03)
Total practice time (years)	11.33 (7.82)	6.00 (2.02)

Means and standard deviations (in parenthesis).

2.2. Ethics

Participants provided written informed consent. The research procedures were approved by the local university ethics committee, following the principles outlined in the Declaration of Helsinki (approval code: CAAE: 85093718.2.0000.5391).

2.3. Tasks, Equipment and Procedures

Balance control was tested in bipedal support, barefoot, keeping the feet hip-width apart with the feet orientated forward in parallel on a force platform (Advanced Mechanical Technology, Inc., model OR6-6, Watertown, MA, USA). Tasks were performed barefoot over a force plate. The evaluation protocol consisted of three tasks, as presented in the following.

(a) Quiet standing. Maintenance of quiet standing, aiming to sustain the motionless upright posture for 30 s (cf. [30]).

(b) Cyclic hip flexion and extension. The initial position was an upright stance with the arms hanging relaxed beside the trunk. The task consisted of performing cyclic hip flexion and extension movements in coordination with shoulder extensions and flexions. In the flexion phase, the hip was flexed about 45 degrees (absolute vertical angle), leaning the trunk forward, while both shoulders were extended up to the arms reaching the horizontal orientation. In the hip extension phase, the reverse movements were performed, with hip extension and shoulder flexion, up to reaching an upright posture with the arms positioned beside the trunk. To favor the reproducibility of movement amplitude, a spatial marker was set individually in front of the participant, at the top of a tripod, at the appropriate height for reaching the specified hip flexion-extension range of motion (Figure 1A,B) (cf. [24]).

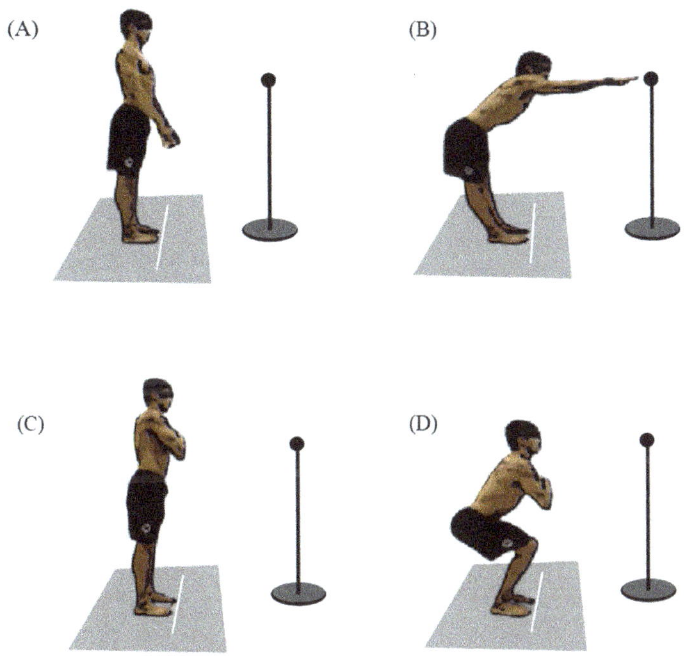

Figure 1. Representation of the postures at the onset and end of each movement phase, for the dynamic balance tasks of hip flexion-extension (**A**,**B**) and squat-lift (**C**,**D**). The top of the vertical shaft served as the spatial reference for standardizing movement amplitude.

(c) Cyclic squat-lift movements. This task emulated sit-to-stand movements employed in functional tests [26,28,31]. The initial position was an upright stance with the arms crossed over the chest. The range of motion was set at about 90° for knee and hip flexion-extension movements. In the squat phase, both the hip and knees were flexed simultaneously, while the trunk was bent forward. A spatial marker was used at the top of a tripod as a reference for the eye's height to finish the squat phase at the desired joint angles. In the lift phase, the reverse movements were performed up to reaching the upright posture. The arms were maintained crossed over the chest throughout a trial (Figure 1C,D).

Both the hip flexion-extension and squat-lift tasks were paced through an electronic metronome (BOSS brand, model DB-60), with trials lasting 20 s. Movement frequency was set at 0.5 Hz, aiming to achieve coincident timing of the end of each movement phase with the metronome beep.

The quiet standing and dynamic balance tasks were tested in the conditions of eyes open and eyes closed. Each task by visual condition was probed over three consecutive trials. Within-task intertrial intervals lasted 15 s, with a 1-min. seated rest interval every three trials. To prevent the after-effects of the dynamic balance tasks, the quiet standing balance was evaluated first. The ensuing sequence of the two dynamic tasks was alternated across participants within the group. For the three tasks, full vision and visual occlusion were alternated between participants within the group.

Immediately preceding the probing trials, participants were familiarized with the task to be performed next. Initially, in the dynamic tasks they assumed the correct maximum hip flexion or squat posture for individually setting the visual marker height and distance. Then, the respective movements were performed in the due range of motion and rhythm, with online feedback provided by an examiner based on subjective online visual evaluation. For the dynamic tasks, the metronome was activated prior to task initiation. This allowed participants to synchronize from the outset their movements with the specified rhythm. In the conditions of visual occlusion, participants were instructed to imagine the location of the visual reference, trying to achieve the specified movement amplitude, and maintaining the head in the vertical orientation, the same way as in the performance under full vision. The performance of the probing trials was visually monitored online by a single examiner (the same across participants). In cases where the performance failed to attend to the required movement amplitude or rhythm, the trial was immediately interrupted. Following extra familiarization movements for stabilization of the required movement characteristics, the testing was reinstated. Interruption occurred in about 2% of trials; no trials were excluded from the analysis.

2.4. Data Collection and Analysis

Ground reaction force data were sampled at a frequency of 200 Hz. After a preliminary visual inspection of individual signals, raw data were processed using MATLAB version 2017b routines (MathWorks, Inc., Natick, MA, USA). Data were digitally filtered using a fourth-order zero-lag Butterworth filter with a cutoff frequency of 10 Hz. The following dependent variables based on center of pressure (center of pressure is a variable frequently used to assess postural stability, representing the point in which the resultant ground reaction force (from the anteroposterior, mediolateral and vertical components) is applied on the support base to sustain quiet stance or dynamic balance) (CoP) displacement were analyzed: peak-to-peak amplitude (delta between the highest and lowest values); root mean square (RMS); and mean velocity. Analyses were conducted separately for the anteroposterior (AP) and mediolateral (ML) directions. For the dynamic tasks, calculations were made for each movement cycle, followed by within trial average over cycles. For the quiet standing and dynamic tasks, variables were calculated for the entire period of task duration. Analysis was based on means from the three trials for each task by visual condition. As a prerequisite for parametric analysis, the Shapiro-Wilk test showed normal data distribution. Analysis was conducted individually for each task through two-way 2 (group: gymnasts X other athletes) X 2 (vision: eyes open X eyes closed) ANOVAs with repeated measures on the last factor. Significant effects ($p < 0.05$) are reported along with the respective effect sizes given by partial eta squared (η_p^2). Statistical analysis was performed using Statistica software (version 7.0, Statsoft, Tulsa, OK, USA). The full dataset is available as Supplementary Material.

3. Results

3.1. Quiet Stance

Figure 2 presents the results of the analysis of CoP displacement amplitude, RMS, and velocity in the AP (panels A–C) and ML (panels D–F) directions. Analysis of CoP sway in the AP direction showed significant main effects of vision. The vision effect was due to higher values for eyes closed compared to eyes open for the three dependent variables: amplitude, $F(1, 17) = 13.66$, $p < 0.01$, $\eta_p^2 = 0.45$; RMS, $F(1, 17) = 6.54$, $p = 0.02$, $\eta_p^2 = 0.28$; and mean velocity, $F(1, 17) = 16.57$, $p < 0.01$, $\eta_p^2 = 0.49$ (Figure 2A–C). Analysis of ML CoP sway showed significant main effects of vision, with higher values for eyes closed compared to eyes open for RMS, $F(1, 17) = 4.71$, $p = 0.04$, $\eta_p^2 = 0.22$; and mean velocity, $F(1, 17) = 12.88$, $p < 0.01$, $\eta_p^2 = 0.43$ (Figure 2D–F). No significant effects related to the group were found for quiet standing.

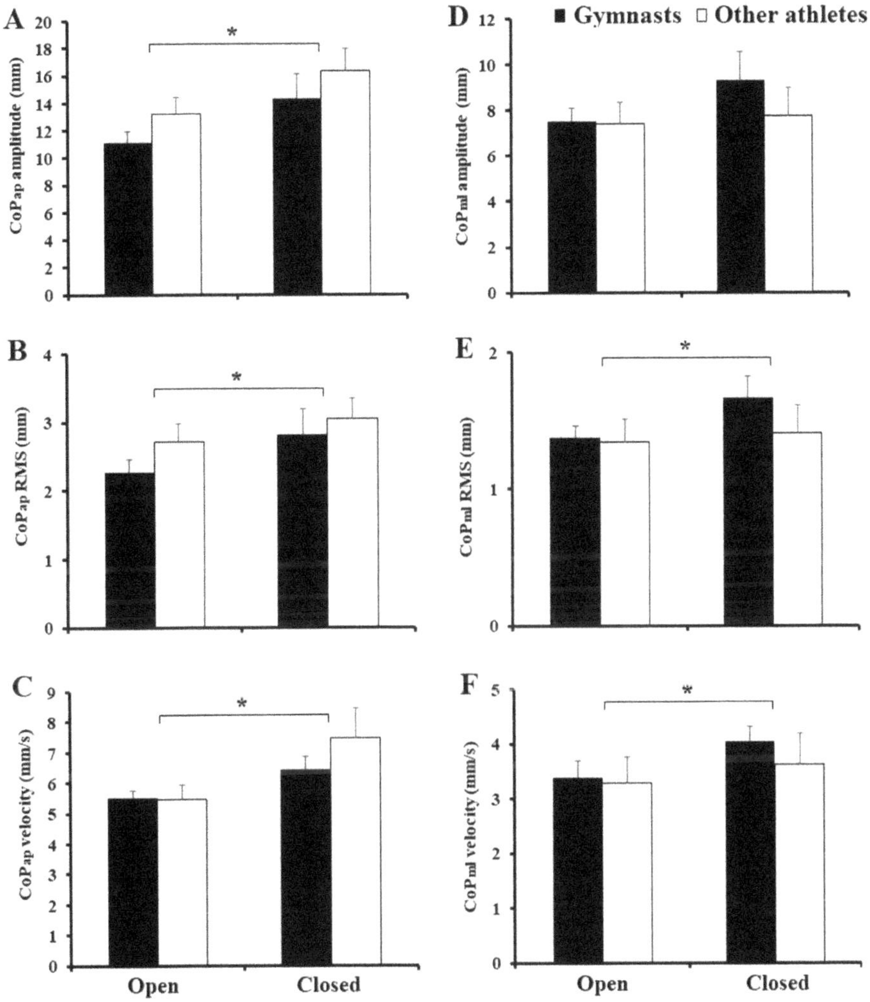

Figure 2. Quiet standing. Comparison between gymnasts and other athletes in the conditions of eyes open and eyes closed; averages (standard deviation indicated by vertical bars) of CoP amplitude (peak-to-peak), CoP root mean square (RMS), and CoP mean velocity in the AP (**A–C**) and ML (**D–F**) directions; significant effects of vision are represented by asterisks.

3.2. Voluntary Dynamic Balance I: Hip Flexion-Extension

Results from CoP analysis for the hip flexion-extension task are presented in Figure 3. Analysis of AP CoP sway showed significantly higher values for eyes closed than eyes open for CoP amplitude, $F(1, 17) = 11.13$, $p < 0.01$, $\eta_p{}^2 = 0.40$; and CoP mean velocity, $F(1, 17) = 23.00$, $p < 0.01$, $\eta_p{}^2 = 0.57$ (Figure 3A–C). Analysis of CoP sway in the ML direction revealed higher values for eyes closed than eyes open for the three CoP-related variables, as follows: amplitude, $F(1, 17) = 50.42$, $p < 0.01$, $\eta_p{}^2 = 0.75$; RMS, $F(1, 17) = 71.95$, $p < 0.01$, $\eta_p{}^2 = 0.81$; and mean velocity, $F(1, 17) = 57.79$, $p < 0.01$, $\eta_p{}^2 = 0.77$ (Figure 3D–F). No significant effects related to the group were found for the hip flexion-extension task.

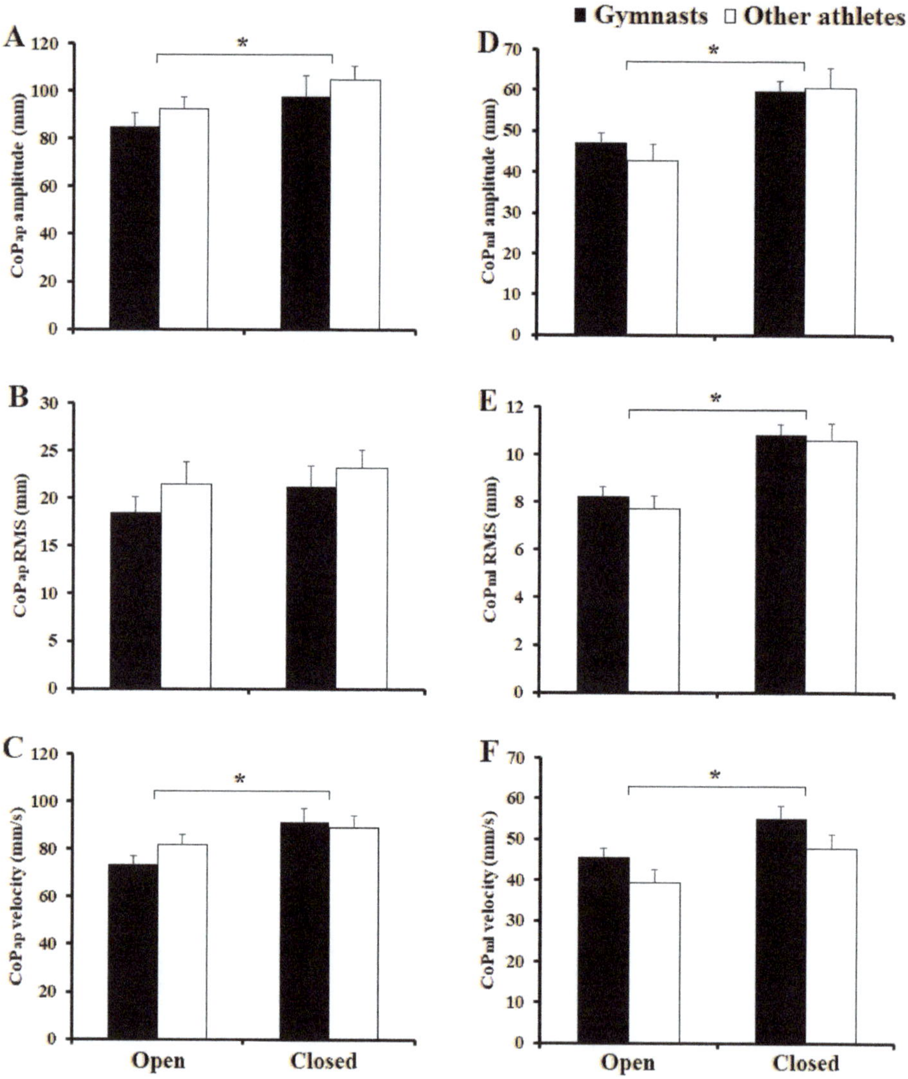

Figure 3. Hip flexion-extension task. Comparison between gymnasts and other athletes in the conditions of eyes open and eyes closed; averages (standard deviation indicated by vertical bars) of CoP amplitude (peak-to-peak), CoP root mean square (RMS), and CoP mean velocity in the AP (**A–C**) and ML (**D–F**) directions; significant effects of vision are represented by asterisks.

3.3. Voluntary Dynamic Balance II: Squat-Lift Task

Analysis of CoP sway for the squat-lift task in the AP direction indicated significant main effects for both the group and vision factors. The group effects were due to lower CoP values in the gymnasts than the athletes from other sports for the three CoP-related variables, as follows: amplitude, $F(1, 17) = 16.42$, $p < 0.01$, $\eta_p{}^2 = 0.49$; RMS, $F(1, 17) = 9.61$, $p < 0.01$, $\eta_p{}^2 = 0.36$; and mean velocity $F(1, 17) = 9.69$, $p < 0.01$, $\eta_p{}^2 = 0.36$. Greater values for the eyes closed than eyes open were found for CoP sway amplitude $F(1, 17) = 6.97$, $p = 0.02$, $\eta_p{}^2 = 0.29$ (Figure 4A–C). Analysis of CoP sway in the ML direction indicated significant main effects of vision. Greater values were found in the eyes closed condition for the three dependent variables, as follows: amplitude, $F(1, 17) = 36.12$, $p < 0.01$, $\eta_p{}^2 = 0.68$; RMS, $F(1, 17) = 41.73$, $p < 0.01$, $\eta_p{}^2 = 0.71$; and mean velocity, $F(1, 17) = 78.48$, $p < 0.01$, $\eta_p{}^2 = 0.82$ (Figure 4D–F).

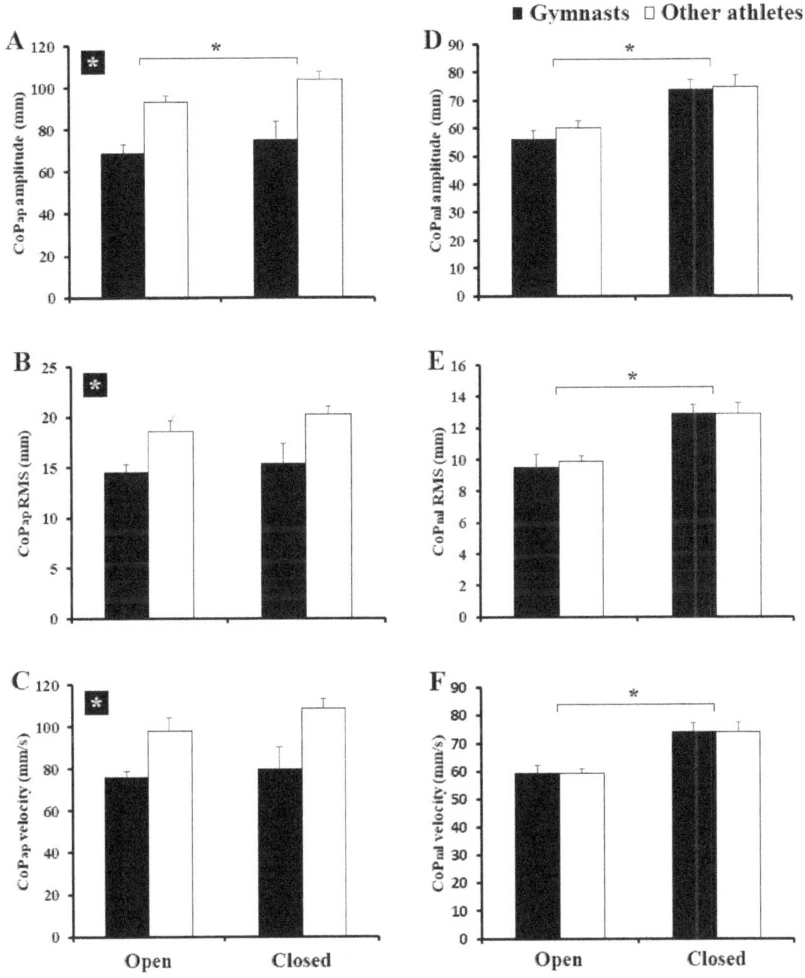

Figure 4. Squat-lift task. Comparison between gymnasts and other athletes in the conditions of eyes open and eyes closed; averages (standard deviation indicated by vertical bars) of CoP amplitude (peak-to-peak), CoP root mean square (RMS), and CoP mean velocity in the AP (**A–C**) and ML (**D–F**) directions; significant effects of vision are represented by black asterisks, while significant effects of group are represented by white asterisks against a black background at the top left corner of the panels.

4. Discussion

In the current investigation, we aimed to analyze the sensitivity of balance evaluation in the performance of quiet standing and two voluntary dynamic tasks by comparing gymnasts and athletes from other sports. The rationale for this comparison is that the expected better balance control in voluntary tasks by gymnasts should be reflected in the performance of tasks with a high demand for voluntary balance. A comparison between the gymnasts and athletes from other sports showed that dynamic balance was task-specific. In the quiet standing and the dynamic hip flexion-extension tasks, no significant differences were found between the gymnasts and the athletes from other sports. Conversely, in the squat-lift task results revealed the expected better performance of the gymnasts over the other athletes, as represented by the reduced amplitude and velocity of CoP displacement during the task execution. Availability of visual information affected the groups similarly, with an equivalent decline of balance stability between the groups in the eyes closed compared to the eyes open condition.

4.1. *Effect of Visual Deprivation*

The effect of deprivation of visual information on body stability is well known in the control of quiet standing, leading to increased amplitude and velocity of balance sway as compared to performance under full vision (e.g., [32,33]), as observed in our results. The effect of vision has also been reported in tasks requiring dynamic balance on an oscillatory support base, with visual occlusion provoking increased amplitudes of head and trunk sway in comparison with performance under full vision [34]. Relevance of visual information has also been shown in reactive balance responses, with visual occlusion leading to a higher velocity of CoP displacement to recover from an extrinsic mechanical stance perturbation [35]. In these balance tasks, visual information is thought to provide a reference of head and trunk stability in space for balance control. In the absence of vision, other sensory sources like the vestibular apparatus [36], plantar cutaneous afferents [37], and proprioceptive receptors [38] may be used as feedback sources for balance control. Our results bring original information on this topic by showing that in both the voluntary hip flexion-extension and squat-lift dynamic tasks amplitude of AP and ML CoP displacement were increased when visual information was suppressed. Lack of vision led to increased CoP velocity in the ML direction for the two dynamic tasks while affecting AP CoP velocity in the hip flexion-extension task only. A point worth noticing is that, differently from other balance tasks, in the execution of the dynamic tasks under examination the head exhibited rapid and continuous movement, encompassing a wide range of motion. This finding suggests that the visual flow, in the focal and/or peripheral vision [39], resulting from voluntary head movements can be employed to stabilize dynamic balance in whole body movements. It seems that the anticipated visual flow resulting from the voluntary head movements, in association with online proprioceptive and plantar cutaneous signals, can be used by the central nervous system for balance regulation while moving. This effect contrasts with the balance perturbation induced by generating a visual flow through a moving room in a quiet stance [40]. In this regard, it can be assumed that the ability to use anticipatory visual flow information in conditions of voluntary head motion is a requirement in our daily living activities, being of paramount importance in the performance of most sports skills.

4.2. *Better Gymnasts' Balance Control in the Squat-Lift Task*

Previous results have shown an effect of task-specificity in the comparison between reactive balance and quiet posture control [9,14]. Specificity in such balance tasks could be explained due to their particular requirements. While quiet standing is regulated through small-scale automatic adjustments to natural body sway, reactive balance responses require identification of the nature, direction and magnitude of an extrinsic stance perturbation and then the generation of a specific response to recover balance stability within a short time interval (cf. [41]). In Ringhof and Stein's [15] investigation, task-specificity was found in gymnasts in a comparison between three tasks, with better gymnasts' balance being detected in a voluntary task requiring a one-legged landing after jumping but not in tasks requiring reactive balance control. While this preliminary study suggested better voluntary balance control in gymnasts specifically for voluntary balance tasks, this result might be due to the extensive practice of gymnasts on landing tasks in their ordinary sport training. Our results revealed that task-specificity can also be seen between two voluntary dynamic tasks, as indicated by better balance performance in gymnasts in the squat-lift but not in the hip flexion-extension task (Figure 3A–C vs. Figure 4A–C). This finding suggests that the squat-lift task was more sensitive to the expected improved voluntary dynamic balance of gymnasts.

A plausible explanation for the increased sensibility of the squat-lift task for balance control evaluation is related to its higher demand for interjoint coordination. The hip flexion-extension task required that the knees were maintained stretched while focal movements were made mainly at the hip. In this action, the hip had to be simultaneously flexed and projected backward to keep the center of mass in a stable position over the support base delimited by the feet support area. This action can be conceived to be relatively simple and overly automatized in movement control. From this perspective, this finding is consistent with previous results showing that better performance of gymnasts over other groups is seen only in tasks imposing higher balance demands given by unipedal stance and malleable support base [19,20]. On the other hand, the squat-lift task involves more complex coordination between the simultaneous motions at the hip and knees to generate the required global movements while preserving balance stability. Our results indicated that the gymnasts had lower values than the athletes from other sports for amplitude and velocity in the AP but not in the ML CoP sway direction. Although balance control in the ML direction has been shown to be associated with performance on the analogous sit-to-stand task in older individuals [28], our results suggest that balance in the frontal plane is insensitive to discriminate interindividual differences of balance control. Supposedly, the demand for symmetric movements between the legs makes the balance demand relatively low in the ML direction for a young sportsperson. We propose that better results of gymnasts than the other athletes in the AP sway direction are due to the high interjoint coordination demand mainly between the hip and knee movements leading to back-and-forth trunk displacements for the squat-lift motion while maintaining the center of mass stably over the support base. Generalizing from upper limb between-joint coordination in reaching actions [42,43], we conceptualize that in the squat-lift task, the central nervous system anticipates and finely regulates through online feedback the interactive torques between the lower limb joints to attenuate the sway magnitude and velocity of the center of mass over the support base. During the cyclic squat-lift movements, dynamic torque variation at the hip, knees and ankles, as well as the reciprocal effects on the adjacent joints, have to be accurately anticipated in the control system to attenuate self-produced balance perturbations by the voluntary movements. An additional point worth noting was that the superior balance performance of gymnasts in the squat-lift task was also observed when vision was suppressed. This finding suggests that the main sensory feedback sources leading to better dynamic balance control in the gymnasts were nonvisual, possibly guided by the myriad of somatosensory signals relevant for balance control which are generated during voluntary whole-body movements (cf. [36–38]).

4.3. Methodological Strengths and Weaknesses

We highlight as the most original methodological advancement in our study the evaluation of voluntary dynamic balance with standardization of movement amplitude and rhythm during the performance of cyclic whole-body movements. With this procedure, we assumedly prevented high intra and interindividual movement variability during the performance of the dynamic tasks as it can occur in protocols in which participants are allowed to perform movements with self-selected spatial and temporal characteristics. This procedure can be thought to favor sustainable conclusions on the interpretation of center pressure values between groups and experimental conditions. The use of gold standard measurements based on the center of pressure provided valid and reliable results in the evaluation of dynamic balance. On the other hand, the lack of kinematic measurements to document the effective amplitude and rhythm of trunk movements across participants represents a limitation in this investigation. It should be acknowledged that performing the tested tasks without vision makes standardizing movement amplitude challenging due to the absence of visual reference. An additional limitation is represented by between-group differences in anthropometric measures (see Table 1). On average, athletes from other sports were approximately 11 cm taller and 16 kg heavier than the gymnasts. It should be noted that anthropometric measures could impact center pressure measurements (cf. [44]).

5. Conclusions and Implications

Our results showed no significant differences in balance control between high-level gymnasts and athletes from different sports for quiet standing and voluntary whole-body movements. The main finding was lower amplitude and velocity displacement of the anteroposterior center of pressure sway in gymnasts compared to athletes from different sports during the voluntary cyclic squat-lift task but not in the hip flexion-extension task. This conclusion was valid for both vision and no-vision conditions. In terms of practical application, these findings suggest that the employed protocol using the squat-lift task could serve as a potentially sensitive method for assessing voluntary dynamic balance. As squat-lifting is a relatively easy task, we speculate that this assessment could apply not only to young individuals but also to older adults, serving as an objective tool for assessing voluntary dynamic balance. To validate this assumption, future studies should incorporate the squat-lift task to evaluate dynamic balance across different age groups.

Supplementary Materials: The following supporting information can be downloaded at: https://www.mdpi.com/article/10.3390/biomechanics4030030/s1, The full dataset is available as supplementary material.

Author Contributions: J.W.F.B., conceptualization, methodology, investigation, analysis, writing—original draft preparation; D.B.C., data curation, analysis, investigation; L.A.T., conceptualization, supervision, writing, review and editing. All authors have read and agreed to the published version of the manuscript.

Funding: This work was supported by the Coordination for the Improvement of Higher Level Personnel (CAPES, grant number 88887.581657/2020-00 [JWFB]), and by the Brazilian Council of Science and Technology (CNPq) under grant number 306323/2019-2 (LAT).

Institutional Review Board Statement: The study was conducted according to the guidelines of the Declaration of Helsinki. Approval number by the local ethics committee: CAAE: 85093718.2.0000.5391.

Informed Consent Statement: Participants provided written informed consent.

Data Availability Statement: The full data set will be available as Supplementary Material.

Acknowledgments: The authors thank Thiago de Camargo e Souza Câmara for their administrative and technical support in conducting this work.

Conflicts of Interest: The authors declare no conflicts of interest. The funders had no role in the design of the study; in the collection, analyses, or interpretation of data; in the writing of the manuscript, or in the decision to publish the results.

References

1. Paillard, T. Plasticity of the postural function to sport and/or motor experience. *Neurosci. Biobehav. Rev.* **2017**, *72*, 129–152. [CrossRef] [PubMed]
2. Valenciano, P.J.; Monteiro, P.H.M.; Lazzaro, I.M.; Heusi da Silva, F.J.M.; Silva, F.H.; Mizrahi, S.E.; Guimarães, C.P.; Teixeira, L.A. Validation of the equidyn protocol for evaluation of dynamic balance in older adults through a smartphone application. *Gait Posture* **2024**, *111*, 59–64. [CrossRef] [PubMed]
3. Butler, R.J.; Southers, C.; Gorman, P.P.; Kiesel, K.B.; Plisky, P.J. Differences in soccer players' dynamic valance across levels of competition. *J. Athl. Train.* **2012**, *47*, 616–620. [CrossRef] [PubMed]

4. Pinheiro, L.S.P.; Ocarino, J.d.M.; Bittencourt, N.F.N.; Souza, T.R.; de Souza Martins, S.C.; Bomtempo, R.A.B.; Resende, R.A. Lower limb kinematics and hip extensors strengths are associated with performance of runners at high risk of injury during the modified Star Excursion Balance Test. *Braz. J. Phys. Ther.* **2020**, *24*, 488–495. [CrossRef] [PubMed]
5. Poncumhak, P.; Srithawong, A.; Duangsanjun, W.; Amput, P. Comparison of the ability of static and dynamic balance tests to determine the risk of falls among older community-dwelling individuals. *J. Funct. Morphol. Kinesiol.* **2023**, *8*, 43. [CrossRef]
6. Nakagawa, T.H.; Petersen, R.S. Relationship of hip and ankle range of motion, trunk muscle endurance with knee valgus and dynamic balance in males. *Phys. Ther. Sport.* **2018**, *34*, 174–179. [CrossRef] [PubMed]
7. Kang, M.-H.; Kim, G.-M.; Kwon, O.-Y.; Weon, J.-H.; Oh, J.-S.; An, D.-H. Relationship between the kinematics of the trunk and lower extremity and performance on the Y-Balance Test. *PMR* **2015**, *7*, 1152–1158. [CrossRef]
8. Schaubert, K.L.; Bohannon, R.W. Reliability and validity of three strength measures obtained from community-dwelling elderly persons. *J. Strength Cond. Res.* **2005**, *19*, 717–720.
9. Rizzato, A.; Bozzato, M.; Zullo, G.; Paoli, A.; Marcolin, G. Center of pressure behavior in response to unexpected base of support shifting: A new objective tool for dynamic balance assessment. *Sensors* **2023**, *23*, 6203. [CrossRef]
10. Coelho, D.B.; Silva, M.B.; de Lima-Pardini, A.C.; Martinelli, A.R.; da Silva Baptista, T.; Ramos, R.T.; Teixeira, L.A. Young and older adults adapt automatic postural responses equivalently to repetitive perturbations but are unable to use predictive cueing to optimize recovery of balance stability. *Neurosci. Lett.* **2018**, *685*, 167–172. [CrossRef]
11. Mochizuki, G.; Boe, S.; Marlin, A.; McIlroy, W.E. Perturbation-evoked cortical activity reflects both the context and consequence of postural instability. *Neuroscience* **2010**, *170*, 599–609. [CrossRef] [PubMed]
12. Teixeira, L.A.; Coutinho, J.d.F.S.; Coelho, D.B. Regulation of dynamic postural control to attend manual steadiness constraints. *J. Neurophysiol.* **2018**, *120*, 693–702. [CrossRef] [PubMed]
13. Rizzato, A.; Gobbi, E.; Paoli, A.; Marcolin, G. Validity and reliability of an unstable board for dynamic balance assessment in young adults. *PLoS ONE* **2023**, *18*, e0280057. [CrossRef]
14. Rizzato, A.; Paoli, A.; Andretta, M.; Vidorin, F.; Marcolin, G. Are static and dynamic postural balance assessments two sides of the same coin? A cross-sectional study in the older adults. *Front. Physiol.* **2021**, *12*, 681370. [CrossRef] [PubMed]
15. Ringhof, S.; Stein, T. Biomechanical assessment of dynamic balance: Specificity of different balance tests. *Hum. Mov. Sci.* **2018**, *58*, 140–147. [CrossRef] [PubMed]
16. Kiers, H.; van Dieën, J.; Dekkers, H.; Wittink, H.; Vanhees, L. A systematic review of the relationship between physical activities in sports or daily life and postural sway in upright stance. *Sports Med.* **2013**, *43*, 1171–1189. [CrossRef] [PubMed]
17. Andreeva, A.; Melnikov, A.; Skvortsov, D.; Akhmerova, K.; Vavaev, A.; Golov, A.; Draugelite, V.; Nikolaev, R.; Chechelnickaia, S.; Zhuk, D.; et al. Postural stability in athletes: The role of sport direction. *Gait Posture* **2021**, *89*, 120–125. [CrossRef] [PubMed]
18. Davlin, C.D. Dynamic balance in high level athletes. *Percept. Mot. Ski.* **2004**, *98*, 1171–1176. [CrossRef] [PubMed]
19. Vuillerme, N.; Danion, F.; Marin, L.; Boyadjian, A.; Prieur, J.M.; Weise, I.; Nougier, V. The effect of expertise in gymnastics on postural control. *Neurosci. Lett.* **2001**, *303*, 83–86. [CrossRef]
20. Vuillerme, N.; Nougier, V. Attentional demand for regulating postural sway: The effect of expertise in gymnastics. *Brain Res. Bull.* **2004**, *63*, 161–165. [CrossRef]
21. Asseman, F.B.; Caron, O.; Crémieux, J. Are there specific conditions for which expertise in gymnastics could have an effect on postural control and performance? *Gait Posture* **2008**, *27*, 76–81. [CrossRef] [PubMed]
22. Gautier, G.; Thouvarecq, R.; Larue, J. Influence of experience on postural control: Effect of expertise in gymnastics. *J. Mot. Behav.* **2008**, *40*, 400–408. [CrossRef] [PubMed]
23. Isableu, B.; Hlavackova, P.; Diot, B.; Vuillerme, N. Regularity of center of pressure trajectories in expert gymnasts during bipedal closed-eyes quiet standing. *Front. Hum. Neurosci.* **2017**, *11*, 317. [CrossRef] [PubMed]
24. Bueno, J.W.F.; Coelho, D.B.; de Souza, C.R.; Teixeira, L.A. Association of foot sole sensibility with quiet and dynamic body balance in morbidly obese women. *Biomechanics* **2021**, *1*, 334–345. [CrossRef]
25. Podsiadlo, D.; Richardson, S. The timed "Up & Go": A test of basic functional mobility for frail elderly persons. *J. Am. Geriatr. Soc.* **1991**, *39*, 142–148. [CrossRef] [PubMed]
26. Jeon, W.; Jensen, J.L.; Griffin, L. Muscle activity and balance control during sit-to-stand across symmetric and asymmetric initial foot positions in healthy adults. *Gait Posture* **2019**, *71*, 138–144. [CrossRef] [PubMed]
27. Lord, S.R.; Murray, S.M.; Chapman, K.; Munro, B.; Tiedemann, A. Sit-to-stand performance depends on sensation, speed, balance, and psychological status in addition to strength in older people. *J. Gerontol. A* **2002**, *57*, M539–M543. [CrossRef] [PubMed]
28. Rezende, L.S.; Monteiro, P.H.; Oliveira, J.A.; Souza, C.R.; Coelho, D.B.; Marcori, A.J.; Teixeira, L.A. Do timed up and go and five times sit to stand test outcomes correlate with trunk stability? A pilot-study. *Braz. J. Mot. Behav.* **2023**, *17*, 150–157. [CrossRef]
29. Schenkman, M.; Hughes, M.A.; Samsa, G.; Studenski, S. The relative importance of strength and balance in chair rise by functionally impaired older individuals. *J. Am. Geriatr. Soc.* **1996**, *44*, 1441–1446. [CrossRef]
30. Termoz, N.; Halliday, S.E.; Winter, D.A.; Frank, J.S.; Patla, A.E.; Prince, F. The control of upright stance in young, elderly and persons with Parkinson's disease. *Gait Posture* **2008**, *27*, 463–470. [CrossRef]
31. Alcazar, J.; Alegre, L.M.; Van Roie, E.; Magalhães, J.P.; Nielsen, B.R.; González-Gross, M.; Júdice, P.B.; Casajús, J.A.; Delecluse, C.; Sardinha, L.B.; et al. Relative sit-to-stand power: Aging trajectories, functionally relevant cut-off points, and normative data in a large european cohort. *J. Cachexia Sarcopenia Muscle* **2021**, *12*, 921–932. [CrossRef] [PubMed]

32. Sarabon, N.; Rosker, J.; Loefler, S.; Kern, H. The effect of vision elimination during quiet stance tasks with different feet positions. *Gait Posture* **2013**, *38*, 708–711. [CrossRef] [PubMed]
33. Sozzi, S.; Monti, A.; De Nunzio, A.M.; Do, M.-C.; Schieppati, M. Sensori-motor integration during stance: Time adaptation of control mechanisms on adding or removing vision. *Hum. Mov. Sci.* **2011**, *30*, 172–189. [CrossRef]
34. Buchanan, J.J.; Horak, F.B. Emergence of postural patterns as a function of vision and translation frequency. *J. Neurophysiol.* **1999**, *81*, 2325–2339. [CrossRef] [PubMed]
35. Martinelli, A.R.; Coelho, D.B.; Teixeira, L.A. Light touch leads to increased stability in quiet and perturbed balance: Equivalent effects between post-stroke and healthy older individuals. *Hum. Mov. Sci.* **2018**, *58*, 268–278. [CrossRef] [PubMed]
36. Horlings, C.G.C.; Küng, U.M.; Bloem, B.R.; Honegger, F.; Van Alfen, N.; Van Engelen, B.G.M.; Allum, J.H.J. Identifying deficits in balance control following vestibular or proprioceptive loss using posturographic analysis of stance tasks. *Clin. Neurophysiol.* **2008**, *119*, 2338–2346. [CrossRef]
37. Meyer, P.F.; Oddsson, L.I.E.; De Luca, C.J. Reduced plantar sensitivity alters postural responses to lateral perturbations of balance. *Exp. Brain Res.* **2004**, *157*, 526–536. [CrossRef] [PubMed]
38. Bloem, B.; Allum, J.H.J.; Carpenter, M.; Verschuuren, J.; Honegger, F. Triggering of balance corrections and compensatory strategies in a patient with total leg proprioceptive loss. *Exp. Brain Res.* **2002**, *142*, 91–107. [CrossRef]
39. Bertenthal, B.I.; Bai, D.L. Infants' sensitivity to optical flow for controlling posture. *Dev. Psychol.* **1989**, *25*, 936–945. [CrossRef]
40. Lee, D.N.; Aronson, E. Visual proprioceptive control of standing in human infants. *Percept. Psychophys.* **1974**, *15*, 529–532. [CrossRef]
41. Teixeira, L.A.; Maia Azzi, N.; de Oliveira, J.Á.; Ribeiro de Souza, C.; da Silva Rezende, L.; Boari Coelho, D. Automatic postural responses are scaled from the association between online feedback and feedforward control. *Eur. J. Neurosci.* **2020**, *51*, 2023–2032. [CrossRef] [PubMed]
42. Maeda, R.S.; Gribble, P.L.; Pruszynski, J.A. Learning new feedforward motor commands based on feedback responses. *Curr. Biol.* **2020**, *30*, 1941–1948. [CrossRef] [PubMed]
43. Maeda, R.S.; Kersten, R.; Pruszynski, J.A. Shared internal models for feedforward and feedback control of arm dynamics in non-human primates. *Eur. J. Neurosci.* **2021**, *53*, 1605–1620. [CrossRef] [PubMed]
44. Hue, O.; Simoneau, M.; Marcotte, J.; Berrigan, F.; Doré, J.; Marceau, P.; Marceau, S.; Tremblay, A.; Teasdale, N. Body weight is a strong predictor of postural stability. *Gait Posture* **2007**, *26*, 32–38. [CrossRef]

MDPI AG
Grosspeteranlage 5
4052 Basel
Switzerland
Tel.: +41 61 683 77 34

Biomechanics Editorial Office
E-mail: biomechanics@mdpi.com
www.mdpi.com/journal/biomechanics

www.ingramcontent.com/pod-product-compliance
Lightning Source LLC
Chambersburg PA
CBHW061629210926
51850CB00009B/2486
* 9 7 8 3 7 2 5 8 5 2 5 5 0 *